AI for Developers

Building Better Software with AI Assistants

Ken Tannenbaum

2026

Finnoybu Press

AI for Developers

2026 Edition

Building Better Software with AI Assistants

Published by Finnoybu Press

For permissions and inquiries: press@finnoybu.org

First Edition: 2026

About the Author

Ken Tannenbaum is a technologist, entrepreneur, and AI governance researcher. He is the founder of Finnoybu Press and the founder/steward of the AEGIS Initiative, an open framework for AI safety and accountability. Learn more at aegis-initiative.com.

When he's not building governance standards for artificial intelligence, he's helping everyday people understand and use it.

Dedication

To every developer who has ever stared at an AI-generated function, thought "that looks right," shipped it, and then spent the next three hours debugging the one edge case the AI confidently ignored.

You are not alone. This book is for you.

(Also, to `node_modules/`. You were the original black box we blindly trusted. AI just made the tradition more conversational.)

The AI for Everyone Series

This book is part of the **AI for Everyone** series by Finnoybu Press, designed to help real people use AI tools effectively. Visit **press.finnoybu.org** for the latest titles and updates.

Collection One: Product Deep Dives

- *ChatGPT: The Definitive Guide (2026 Edition)*
- *ChatGPT: The Power User Guide (2026 Edition)*
- *Claude: The Definitive Guide (2026 Edition)*
- *Claude Code: The Power User Guide (2026 Edition)*
- *Gemini: The Complete Guide (2026 Edition)*
- *Gemini: The Advanced Guide (2026 Edition)*
- *Copilot: The Workplace AI Guide (2026 Edition)*
- *Copilot: The Advanced Workplace Guide (2026 Edition)*
- *Perplexity: The Research Playbook (2026 Edition)*
- *Perplexity: The Advanced Research Playbook (2026 Edition)*

Collection Two: Product Guides

- *NotebookLM: From Sources to Audio Overviews (2026 Edition)*
- *DeepSeek: Code, Math, and Reasoning (2026 Edition)*
- *Cursor: The AI Code Editor (2026 Edition)*
- *OpenClaw: Agentic AI for Everyday Work (2026 Edition)*

Collection Three: Cross-Platform Skills

- *AI Systems Playbook (2026 Edition)*
- *AI Prompt Engineering (2026 Edition)*
- *AI for Developers (2026 Edition)*
- *AI for Business (2026 Edition)*
- *AI for Students (2026 Edition)*
- *AI Governance for Practitioners (2026 Edition)*
- *Securing Agentic AI (2026 Edition)*

Table of Contents

Introduction

What This Book Is

This is the book I wish existed when I started using AI to write code.

Not a product manual. Not a prompt cookbook. Not a manifesto about how AI will replace developers (it will not) or a screed about how AI is just autocomplete (it is not).

This is a practical, comprehensive guide to integrating AI into your software development workflow — across every major tool, every phase of development, and every type of project you are likely to encounter. It covers Claude Code, GitHub Copilot, Cursor, ChatGPT, Gemini Code Assist, Amazon Q Developer, Windsurf, and others. It teaches you how to use them together, when to reach for each one, and — critically — when to close the AI tab and think for yourself.

If you have read other books in the AI for Everyone series, you know the format: practical, opinionated, example-heavy, honest about limits. If this is your first, welcome. You are in the right place.

Who This Book Is For

You are a software developer. You can already code. You do not need this book to explain what a function is, how a REST API works, or why you should use version control.

You might be:

- **A professional developer** who has dabbled with AI tools but lacks a systematic workflow around them
- **A senior engineer** evaluating AI tools for your team and wanting a clear-eyed assessment
- **A full-stack developer** who uses Copilot for autocomplete but suspects there is more to the story
- **A solo developer** building products and wanting to multiply your output without multiplying your bugs

- **A student or bootcamp graduate** who codes competently and wants to level up with AI
- **A tech lead** establishing AI coding practices for a team

This book assumes you write code regularly in at least one language, understand basic software engineering concepts (testing, version control, deployment), and have a healthy skepticism about tools that promise to do your job for you.

What This Book Is NOT

This is not a beginner's guide to programming. If you cannot read a Python function or a TypeScript interface, start with a programming book first.

This is not a guide to building AI/ML models. If you want to train neural networks, this is the wrong shelf. Chapter 22 covers adding AI features to your applications via APIs, but the focus is using AI as a development tool, not building AI itself.

This is not a product-specific deep dive. For that level of detail on individual tools, see the product guides in Collections 1 and 2 of this series. This book covers the landscape, compares tools, and teaches cross-tool workflows.

How This Book Is Organized

The book is divided into five parts:

Part I: Foundations (Chapters 1–3) gets you oriented. What AI means for development, what tools exist, and how to set up an AI-augmented development environment.

Part II: Core Development Skills (Chapters 4–9) covers the daily tasks every developer does — writing code, reviewing code, debugging, refactoring, testing, and documentation — with AI assistance.

Part III: Domain-Specific Development (Chapters 10–16) goes deep on architecture, frontend, backend, DevOps, large codebases, git workflows, and API development.

Part IV: Advanced Concerns (Chapters 17–22) tackles the hard stuff — security, databases, mobile development, pair programming workflows, failure modes, and building AI-powered features into your own applications.

Part V: The Bigger Picture (Chapters 23–24) covers team adoption and the future of AI-assisted development.

The appendices provide reference material: tool comparisons, prompt patterns, configuration guides, a glossary, and further reading.

How to Read This Book

You have two good options:

Option A: Front to back. The chapters build on each other. Part I gives you the mental models, Part II gives you the core skills, and Parts III-IV assume you have internalized both.

Option B: Jump to what you need. Every chapter stands on its own. If you are debugging a production issue right now and want AI help, skip to Chapter 6. If you are evaluating tools for your team, start with Chapter 2 and Appendix A. Cross-references point you to prerequisite material when it matters.

A Note About Accuracy

The AI landscape changes constantly. New models launch monthly, features get added, prices shift, best practices evolve. This book is accurate as of May 2026, but specific details — vendor pricing, feature availability, exact tool capabilities — may have shifted by the time you read this.

The principles, techniques, and strategies in this book have a much longer shelf life than specific tool details. How to prompt effectively for code, verify AI-generated output, and structure agentic development workflows — those skills transfer across versions and across different AI tools.

When something seems different from what's described here, check the vendor's documentation for the latest information.

Conventions Used in This Book

Code examples appear in fenced code blocks with language annotations:

```python
def greet(name: str) -> str:
    return f"Hello, {name}!"
```

When I reference a specific AI tool's output, I label it clearly:

> **Claude:** Here is a function that handles the edge case...
>
> **Copilot suggestion:** (appears as ghost text in your editor)

Tool-specific tips appear in callout boxes labeled with the tool name. Cross-tool advice — the stuff that works everywhere — appears as regular text.

Keyboard shortcuts follow the format `Ctrl+K` (Windows/Linux) or `Cmd+K` (macOS), Windows first.

Let's Get Started

The AI revolution in software development is not coming. It is here. You are late to the party, but the party is just getting interesting.

Open your editor. Pick your AI tool. Let's build something.

Part I: Getting Started

"Talk is cheap. Show me the code."

— Linus Torvalds

Chapter 1: The AI-Augmented Developer

How AI Changes (and Doesn't Change) Software Development

Every developer has the moment the first time they use a good AI coding tool. You describe a problem in plain English, and the AI writes a function that works. Not sort-of works. Works. The tests pass. The edge cases are handled. The code is cleaner than what you would have written at 4 PM on a Friday.

And then a quiet voice in the back of your head whispers: "Am I about to be replaced?"

No. But you are about to be different.

Augmentation, Not Replacement

Clear the existential crisis out of the way first, because it poisons every conversation about AI in development until you resolve it.

AI coding tools in 2026 are extraordinarily capable at generating code from well-specified requirements. They are equally capable at generating confident, plausible, subtly wrong code from vague requirements. The difference between those two outcomes is you — the developer who knows what to ask for, how to verify the output, and when the AI is hallucinating an API that does not exist.

Here is what AI does better than you:

- **Boilerplate generation.** Express route handlers, React component scaffolding, database migration files, Terraform configurations. Anything that follows a well-established pattern, AI writes faster and more consistently than you do.
- **Language translation.** Converting a Python algorithm to Go, a SQL query to a different dialect, a REST API call to GraphQL. AI handles syntactic translation with remarkable accuracy.

- **Pattern recognition in errors.** Given a stack trace and relevant code, AI often identifies the root cause faster than you, because it has seen thousands of variations of every common error.
- **Documentation generation.** Docstrings, README sections, API documentation, and inline comments from existing code. AI is tireless and thorough in ways humans are not.

Here is what AI does worse than you:

- **Understanding your business domain.** AI does not know your e-commerce platform treats "out of stock" differently from "discontinued" and that confusing them will break downstream fulfillment. You do.
- **System-level reasoning.** AI can write a perfect function that is the wrong function. It optimizes locally. You think systemically about how this code interacts with everything else.
- **Judgment under ambiguity.** When requirements conflict, when there is no "right" answer, when you need to trade off performance against maintainability — that is human territory.
- **Knowing when to say no.** AI will happily implement any feature you describe, including the ones that should never be built. You push back on bad product decisions.

The AI-augmented developer is not a developer who types less. It is a developer who thinks more. You spend less time on mechanical coding and more time on architecture, design, verification, and the parts of software engineering that require engineering.

The Productivity Multiplier (With Caveats)

You have probably seen the studies. GitHub's research claims Copilot makes developers 55% faster. Various companies report 30–40% productivity gains. These numbers are real, but they require context.

The productivity gain is not uniform. It depends on:

The type of work. A new CRUD API endpoint with AI assistance? Enormous speedup. Debugging a race condition in a distributed system? AI might help you reach the root cause faster, but the deep thinking is still on you.

Your prompting skill. A developer who writes "make a login form" gets mediocre results. A developer who writes "Create a React login form component using React Hook Form with Zod validation, email and password fields, error display, loading state, and an onSubmit callback that accepts validated credentials" gets production-ready code. The gap between those two prompts is the gap between a 10% speedup and a 60% speedup.

Your verification discipline. If you accept AI output without review, you will ship bugs faster. Congratulations, you are now 55% faster at creating technical debt. Real productivity means the code works, is maintainable, and does not introduce security vulnerabilities.

The codebase context. AI tools that understand your project — Claude Code reading your codebase, Copilot trained on your repo's patterns — produce dramatically better results than context-free generation. Chapter 14 covers this in depth.

Here is an honest assessment of where the speedup lands for a competent developer who uses AI well:

Task	Speedup	Notes
Boilerplate / scaffolding	5–10x	AI's strongest suit
Implementing well-defined features	2–3x	Requires good prompting
Writing tests	2–4x	Especially for edge cases
Documentation	3–5x	Tireless and thorough
Debugging	1.5–2x	Depends heavily on context
Architecture / design	1.2–1.5x	AI helps explore options
Complex algorithmic work	1–1.5x	Minimal improvement
Novel problem solving	0.8–1.2x	Can actually slow you down

Notice that last row. For novel problems — the kind where you do not know what you are building until you build it — AI can slow you down. You spend time prompting, reading output, and course-correcting when you would have been better off thinking with a whiteboard.

The New Developer Skill Stack

Being an AI-augmented developer requires new skills layered on top of your existing ones. Think of it as an upgrade, not a replacement.

Skill 1: Prompt Engineering for Code

The obvious one, covered in depth throughout this book. Short version: specify what you want precisely enough that AI can deliver it, while leaving enough flexibility that AI can apply patterns you might not have considered.

Bad prompt:

```
Write a user service
```

Good prompt:

```
Write a UserService class in TypeScript that:
- Uses a PostgreSQL database via Prisma ORM
- Implements CRUD operations (create, getById, getByEmail, update, delete)
- Soft-deletes users (sets deletedAt timestamp instead of removing rows)
- Throws typed errors (UserNotFoundError, DuplicateEmailError)
- All methods are async and return typed results
- Include JSDoc comments on all public methods
```

The good prompt takes 30 seconds longer to write and saves you 20 minutes of editing the output.

Skill 2: AI Output Verification

The less obvious but more important skill. Read AI-generated code with the same critical eye you would apply to a pull request from a junior developer — one who is very smart, very fast, and occasionally very wrong.

Verification means:

- **Reading every line.** Do not skim. The bug is in the line you skipped.
- **Checking imports and dependencies.** AI regularly suggests packages that do not exist, or uses the wrong version's API.
- **Testing edge cases.** AI code typically nails the happy path and fumbles the edges.

- **Verifying security.** AI-generated code routinely has SQL injection vulnerabilities, missing input validation, and exposed secrets. Chapter 17 covers this in detail.

Skill 3: Context Management

AI tools produce dramatically better output with the right context. That means learning to:

- Structure your project so AI can understand it (clear file organization, good naming, CLAUDE.md or similar project files)
- Feed relevant context without overwhelming the model (the right files, not all files)
- Use tool-specific context features (Copilot's `@workspace`, Claude Code's codebase awareness, Cursor's `@codebase`)

Skill 4: Multi-Tool Orchestration

No single AI tool is best at everything. The best developers in 2026 use multiple tools:

- **GitHub Copilot** for inline autocomplete while typing
- **Claude Code** for complex, multi-file changes and codebase-wide operations
- **ChatGPT or Claude** for architecture discussions and design exploration
- **Cursor** for AI-native editor workflows
- **Specialized tools** for specific domains (database tools, security scanners, etc.)

Learning when to reach for which tool is a skill this book will teach you.

Skill 5: Knowing When to Stop

This might be the most important skill of all. There are moments when AI is not helping:

- When you are on your third iteration of a prompt and the output keeps missing the point
- When you are debugging AI-generated code you do not fully understand

- When you are stacking complexity onto your prompt to work around AI limits instead of writing the code yourself
- When the AI is confidently wrong and you are starting to doubt your own correct understanding

In these moments, close the AI tool. Open a blank file. Write the code yourself. You are still a developer. The AI is a tool, not a crutch.

What Actually Changed

Concretely, what is different about day-to-day development work in 2026 compared to 2023.

Code reviews changed. You now review AI-generated code as much as human-written code. The criteria are different — AI code tends to be syntactically clean but semantically subtle in its bugs. You look for different things.

Documentation improved. For the first time in the industry's history, documentation is getting written. Not because developers suddenly started caring about docs, but because generating decent documentation from existing code takes 30 seconds instead of 30 minutes.

Testing got more thorough. AI-generated test suites cover edge cases developers would not think of. Not because AI is smarter, but because generating 50 test cases costs the same effort as generating 5.

Architecture conversations got richer. Instead of one architect proposing a design in a meeting, you can generate and compare multiple architectural approaches in an hour. "Here's a microservices approach, here's a modular monolith approach, here's the trade-off analysis" — AI makes it practical to explore options before committing.

Onboarding accelerated. New developers can ask AI to explain unfamiliar codebases, generate examples of how internal APIs are used, and ramp up dramatically faster. This is one of AI's most underappreciated benefits.

The bar for "good enough" code rose. When boilerplate is free, there is no excuse for inconsistent error handling, missing input validation, or sloppy structure. The team can demand higher standards because meeting them is no longer expensive.

What Did Not Change

Some things did not change, and it is worth being clear about them.

Software is still hard. The hard parts of software engineering — understanding requirements, managing complexity, making architectural decisions, coordinating teams — are not automated by AI. They might be assisted, but the thinking is still yours.

Debugging is still debugging. AI helps you find bugs faster, but the bugs are still there. AI also introduces new categories of bugs that did not exist before (see Chapter 21).

Technical judgment still matters. Knowing that a particular approach will not scale, that a dependency is poorly maintained, that a design pattern is wrong for this context — these judgments come from experience, not from AI.

You still need to understand the code. If you ship code you do not understand, you will pay for it eventually. AI-generated or not, every line in your codebase is your responsibility.

The AI-Augmented Developer Mindset

Here is the mental model to carry through the rest of this book:

You are the architect. AI is a very fast, very knowledgeable, somewhat unreliable contractor.

You would not let a contractor design your house. But you would let a contractor frame walls, run wiring, and pour concrete — as long as you reviewed the work and verified it met spec.

That is what AI-augmented development looks like. You design. You specify. You review. You verify. AI executes the mechanical parts at 10x speed. You apply judgment at human speed, because judgment is the part that matters.

The developers who thrive in this era are the ones who get this balance right. Not the ones who blindly accept AI output. Not the ones who refuse to use AI out of pride. The pragmatists who use every tool available and apply their own judgment to every decision.

That is what this book teaches. Let's get into the specifics.

Chapter 2: The AI Coding Tool Landscape

Claude Code, GitHub Copilot, Cursor, ChatGPT, Gemini Code Assist, Amazon Q, and More

The AI coding tool market in 2026 is a crowded, noisy, rapidly shifting landscape. New tools launch monthly. Existing tools ship major features weekly. Marketing claims are breathless and often misleading. Cutting through the noise to see what each tool actually does well — and where it falls short — is essential before you commit to a workflow.

This chapter gives you that assessment. Honest, practical, based on real usage. We cover every major tool, what it excels at, where it falls short, and who should use it. Appendix A has the detailed comparison matrix; this chapter tells the story.

The Categories

AI coding tools fall into three broad categories, and understanding the categories matters more than memorizing feature lists:

Inline Completion Tools sit inside your editor and suggest code as you type. They predict what you are about to write and offer to complete it. GitHub Copilot pioneered this category. The interaction model is passive — you code, the AI suggests, you accept or reject.

Chat-Based Assistants are conversational tools you interact with by describing what you want. ChatGPT, Claude (via the web or API), and Gemini fall here. You ask questions, describe problems, get code in response. The interaction model is active — you prompt, the AI responds.

Agentic Development Tools go further. They read your codebase, execute commands, make multi-file changes, run tests, and iterate on their output. Claude Code, Cursor's agent mode, and Amazon Q Developer's agent features belong here. The interaction model is collaborative — you describe the goal, the AI works toward it semi-autonomously.

Most modern tools blur these categories. Copilot has chat. Claude has an API you can use for inline completion. Cursor has both completions and agentic capabilities. But the core interaction model still determines how you use each tool day-to-day.

GitHub Copilot

What it is: An AI coding assistant developed by GitHub (Microsoft) and powered primarily by OpenAI models. Available as a VS Code extension, JetBrains plugin, and in the GitHub.com interface.

What it does best:

Copilot's inline completion is the gold standard. When Copilot suggests the next line, or the next function, or fills in a switch statement — the experience is uncanny. It is trained on a massive corpus of code and excels at predicting patterns.

Copilot is particularly strong at: - Completing function bodies from signatures and docstrings - Generating repetitive code patterns (switch statements, error handling blocks) - Writing code that follows the conventions of your current file - Suggesting test cases when you start writing a test function

The tight VS Code integration means it feels native. Tab to accept. `Alt+]` to cycle through suggestions. No context switching.

Where it falls short:

Copilot's completions are often syntactically correct but logically wrong in subtle ways. It optimizes for what looks right, not what is right. It has limited project-wide context — it sees the current file and open tabs, but does not deeply understand your architecture.

Copilot Chat improved things significantly, but it still lacks the deep codebase understanding that tools like Claude Code provide. For complex, multi-file refactoring, Copilot struggles.

Pricing (as of early 2026): - Free tier: Limited completions per month - Individual: $10/month - Business: $19/user/month - Enterprise: $39/user/month

Best for: Developers who want inline autocomplete while typing. Nearly every developer should have Copilot (or a competitor) active in their editor. The productivity gain on routine code is too large to ignore.

```
## Copilot excels at completing patterns like this.
## You write the function signature and docstring:

def calculate_shipping_cost(
    weight_kg: float,
    destination_country: str,
    is_express: bool = False
) -> Decimal:
    """Calculate shipping cost based on weight, destination, and speed.

    Domestic: $5 base + $2/kg, express adds 50%
    International: $15 base + $5/kg, express adds 75%
    """
    # Copilot will reliably complete this entire function body
    # from the signature and docstring above.
```

Claude Code

What it is: Anthropic's CLI-based agentic coding tool. It runs in your terminal, reads your entire codebase, executes commands, and makes changes across multiple files.

What it does best:

Claude Code is the most capable tool for complex, codebase-aware development tasks in 2026. Its killer feature is deep project understanding — it reads your files, understands your architecture, follows your conventions, and makes changes consistent with your existing code.

It excels at: - Multi-file refactoring (rename a concept across your entire project) - Complex feature implementation that touches multiple layers - Codebase exploration and understanding - Debugging with full project context - Writing tests that match your testing patterns - Creating and reviewing pull requests

Claude Code's `CLAUDE.md` project files give it persistent context about your project's conventions, architecture, and preferences. Most other tools lack this.

Where it falls short:

It is a CLI tool. If you want a GUI, use it through an editor integration (VS Code extension) or a wrapper. The terminal-first approach is powerful but has a learning curve.

It is slower than inline completion tools. You describe what you want, wait for it to think, then review the output. For quick "finish this line" work, Copilot is faster.

Claude Code can be expensive for heavy usage due to API token costs, and the model's context window, while large, still has limits with very large codebases.

Pricing: Claude Code offers both subscription and API-based pricing. The Max 5x ($100/month) and Max 20x ($200/month) plans are both Max-tier individual subscriptions, differentiated by usage allotment, providing a fixed monthly cost with generous limits. Alternatively, you can pay based on Anthropic API usage (token-based), where typical cost is $5–50/day for active development depending on usage. The subscription is simpler to budget; the API is more cost-effective for lighter usage.

Best for: Complex development tasks, multi-file changes, working with large codebases, and developers who prefer terminal workflows. This is the tool for heavy lifting.

```
## Claude Code in action — describe what you want, it does the rest
$ claude "Refactor the authentication module to use JWT instead of
  session tokens. Update the middleware, user routes, and tests."

## It reads your codebase, plans the changes, modifies files,
## runs tests, and presents the results for your review.
```

Cursor

What it is: An AI-native code editor (fork of VS Code) built around AI interaction. It combines inline completion, chat, and agentic capabilities in a single editor.

What it does best:

Cursor's pitch is integration. Instead of bolting AI onto an editor, it builds the editor around AI. The `Cmd+K` inline editing, `Cmd+L` chat, and agent mode create a fluid workflow where AI is always a keystroke away.

Standout features: - **Inline editing**: Select code, press `Cmd+K`, describe what you want changed, and it rewrites the selection - **Codebase-aware chat**: `@codebase` lets the chat understand your entire project - **Multi-file editing**: Agent mode can create, modify, and delete files across your project - **Tab completion**: Copilot-style suggestions with its own model

Cursor is popular with frontend developers because it excels at generating React components, CSS, and UI code from descriptions.

Where it falls short:

You have to use the Cursor editor. If your team standardizes on VS Code, JetBrains, or Vim, adding a second editor creates friction. Cursor is VS Code-based, so extensions mostly work, but it is a separate tool with its own update cycle and occasional compatibility issues.

Output quality depends on which model you select (Claude, GPT-4, etc.), and not all models are available on all plans.

Pricing (as of early 2026): - Hobby: Free - Pro: $20/month - Pro+: $60/month - Ultra: $200/month - Teams: $40/user/month - Enterprise: Custom

Best for: Developers who want AI deeply integrated into their editor experience and are willing to switch (or add) editors. Excellent for frontend work and rapid prototyping.

ChatGPT

What it is: OpenAI's general-purpose AI assistant, accessible via web interface, desktop apps, and API. Not specifically a coding tool, but widely used for development.

What it does best:

ChatGPT is the Swiss Army knife. Not specialized for coding, but its breadth of knowledge makes it valuable for:

- **Exploring architectural options**: "Compare event sourcing vs. CQRS for this use case"

- **Learning new technologies**: "Explain Kubernetes networking to someone who understands Docker"
- **Algorithm design**: "What is the most efficient approach for finding overlapping intervals?"
- **Code explanation**: "What does this regex do?" (with pasted code)
- **Cross-domain knowledge**: When your coding problem intersects with domain knowledge (finance, healthcare, legal requirements), ChatGPT's breadth is an advantage

ChatGPT's canvas feature enables collaborative code editing in the browser, useful for iterating on algorithms or exploring approaches before committing to implementation.

Where it falls short:

ChatGPT has no direct access to your codebase. You copy-paste code into the chat, which limits the context it can work with. For anything involving multiple files or project-wide understanding, this is a significant limit.

It also tends to generate code that is correct in isolation but does not match your project's conventions, patterns, or dependencies. You spend time adapting its output to fit your codebase.

Pricing: - Free tier: GPT-5.4 mini, limited usage - Plus: $20/month - Pro: $200/month - Team: $30/user/month

Best for: Architecture discussions, learning new technologies, exploring approaches, and one-off code problems. Use it as a thinking partner, not as a code generator for production.

Gemini Code Assist

What it is: Google's AI coding assistant, integrated into VS Code, JetBrains IDEs, and Google Cloud. Powered by Gemini models.

What it does best:

Gemini Code Assist has deep integration with Google Cloud services and the Google development ecosystem. If you are building on GCP, Firebase, or Android, Gemini understands those platforms natively.

Strong areas: - **Google Cloud integration**: Generates GCP-specific code (Cloud Functions, BigQuery queries, Pub/Sub handlers) with correct configurations - **Android development**: Deep grasp of Android APIs, Kotlin, and Jetpack Compose - **Large context windows**: Gemini's 1M+ token context window ingests large files and codebases - **Code search**: Understands your repository structure and answers questions about it

Where it falls short:

Outside the Google ecosystem, Gemini Code Assist is less differentiated. Its inline completions are competent but not as refined as Copilot's. Its agentic capabilities are less mature than Claude Code's or Cursor's agent mode.

Pricing: - Free tier within Google Cloud - Enterprise: Part of Gemini for Google Cloud pricing

Best for: Teams building on Google Cloud Platform, Android developers, and organizations already invested in the Google ecosystem.

Amazon Q Developer

What it is: AWS's AI coding assistant (formerly CodeWhisperer). Integrated into VS Code, JetBrains IDEs, and AWS services.

What it does best:

Amazon Q Developer knows AWS. If you are writing Lambda functions, configuring DynamoDB tables, setting up API Gateway routes, or writing CDK infrastructure code, Q Developer generates correct, idiomatic code that follows AWS best practices.

Key strengths: - **AWS service integration**: Generates correct IAM policies, CloudFormation templates, and CDK code - **Security scanning**: Built-in scanning catches common vulnerabilities in AI-generated and human-written code - **Code transformation**: Upgrades Java applications (e.g., Java 8 to 17) and .NET applications - **Agent capabilities**: Implements features, generates tests, and handles multi-step tasks within AWS projects

Where it falls short:

Outside of AWS, Q Developer is less compelling. Its general-purpose coding capabilities lag Copilot, Claude Code, and Cursor. If you are not building on AWS, there is little reason to choose Q Developer as your primary tool.

Pricing: - Free tier: Individual developers (limited) - Pro: $19/user/month - Part of broader AWS enterprise agreements

Best for: Teams building on AWS. Essential for infrastructure code and Lambda development. Use alongside a general-purpose tool like Copilot.

Windsurf (by Codeium)

What it is: An AI-native code editor built by Codeium, similar in concept to Cursor but with its own model stack and "Flows" feature for multi-step workflows.

What it does best:

Windsurf's "Cascade" agent handles multi-step development tasks — implementing features, fixing bugs, writing tests — with a focus on following your codebase patterns. Its "Flows" system tracks what you are working on and proactively suggests next steps.

Where it falls short:

Windsurf is newer and has a smaller community than Cursor. Extension compatibility can be inconsistent. The "Flows" feature, while interesting, sometimes suggests irrelevant actions.

Pricing: - Free tier available - Pro: $15/month - Enterprise: Custom pricing

Best for: Developers who like the AI-native editor concept but want an alternative to Cursor, or who find Windsurf's pricing more attractive.

Other Notable Tools

JetBrains AI Assistant: Built into IntelliJ IDEA, PyCharm, WebStorm, and other JetBrains IDEs. Uses multiple models (including Anthropic and OpenAI). Best for developers committed to JetBrains who want native AI without switching editors.

Tabnine: Focuses on code completion with enterprise-grade privacy. Runs models locally if needed. Best for organizations with strict data privacy requirements that cannot send code to external APIs.

Sourcegraph Cody: AI assistant with deep code search integration. Excels at understanding large codebases and answering questions about code. Best for teams with large monorepos who need to navigate complex code.

Replit AI: Integrated into the Replit browser-based IDE. Strong for prototyping and educational projects. Less suited for production development workflows.

Aider: Open-source, terminal-based AI coding assistant. Similar in concept to Claude Code but model-agnostic. Best for developers who want full control over which models they use.

How to Choose

You do not need to pick one tool. Most productive developers in 2026 use at least two:

1. **An inline completion tool** (Copilot, Cursor's tab completion, Gemini Code Assist) — always on while you type
2. **A deep reasoning tool** (Claude Code, Cursor agent mode, ChatGPT for architecture) — for complex tasks

If you are building on a specific cloud platform, add the platform-specific tool: - AWS → Amazon Q Developer - GCP → Gemini Code Assist - Azure → GitHub Copilot (tight Microsoft integration)

My recommendation for most developers:

Role	Primary Stack
Full-stack web developer	Copilot + Claude Code
Frontend specialist	Cursor (all-in-one)
Backend/systems developer	Copilot + Claude Code
Cloud/DevOps engineer	Copilot + platform-specific tool
Mobile developer	Copilot + Gemini (Android) or Copilot + Claude (iOS)

Role	Primary Stack
Solo developer / indie hacker	Cursor or Copilot + Claude Code
Enterprise team	Copilot Enterprise + Claude Code or Cursor Business

The landscape will keep evolving. The principles in this book — how to prompt effectively, verify output, and structure your workflow — apply regardless of which tool you use. Tools change. Skills transfer.

A Word on Cost

AI coding tools are not free. Even the "free tiers" are limited enough that serious developers hit the ceiling quickly.

Budget $20–60/month for AI tooling. That is less than most developers spend on coffee, and the productivity return is dramatically higher. If your employer does not cover AI tooling costs, make the case. The math is obvious: even a 10% productivity improvement for a developer earning $150K/year saves $15K/year. A $240–720/year tool investment is a rounding error.

If you are a solo developer or student, start with free tiers. Copilot's free tier plus Claude's free tier gives you a solid starting point. Upgrade when the free tiers start limiting your productivity.

Chapter 3: Setting Up Your AI Development Environment

Installing and Configuring Multiple AI Tools for Maximum Effectiveness

A developer with one AI tool is productive. A developer with three AI tools configured to work together is dangerous — in the best way.

This chapter walks you through setting up a complete AI-augmented development environment. Not just installing tools, but configuring them to complement each other, setting up project-level AI context files, and building the habits that make AI assistance feel natural rather than disruptive.

The Target Setup

By the end of this chapter, you will have:

1. An inline completion tool running in your editor (GitHub Copilot or equivalent)
2. An agentic coding tool in your terminal (Claude Code)
3. A chat-based assistant on hand for thinking through problems (Claude, ChatGPT, or Gemini)
4. Project-level configuration files that give your AI tools context about your codebase
5. Keyboard shortcuts that make switching between AI tools frictionless

This setup covers 95% of AI-assisted development scenarios. You can add specialized tools later, but this foundation handles daily work.

Installing GitHub Copilot

Copilot is an extension install in VS Code or a plugin in JetBrains. The setup is straightforward.

VS Code

1. Open VS Code
2. Go to Extensions (`Ctrl+Shift+X`)
3. Search for "GitHub Copilot"
4. Install both "GitHub Copilot" and "GitHub Copilot Chat"
5. Sign in with your GitHub account when prompted
6. Verify: open any code file and start typing — you should see gray ghost text suggestions

JetBrains (IntelliJ, PyCharm, WebStorm, etc.)

1. Go to Settings → Plugins
2. Search for "GitHub Copilot"
3. Install and restart the IDE
4. Sign in with your GitHub account via the toolbar icon

Configuration Worth Changing

The defaults work for most developers, but consider these adjustments:

```
// VS Code settings.json
{
  // Show completions inline (default: true, keep it)
  "github.copilot.enable": {
    "*": true,
    "plaintext": false,  // Disable for plain text files
    "markdown": true,     // Keep for markdown (useful for docs)
    "yaml": true          // Keep for config files
  },

  // Copilot Chat settings
  "github.copilot.chat.localeOverride": "en",

  // Control when suggestions appear
  "editor.inlineSuggest.enabled": true,
  "editor.inlineSuggest.showToolbar": "onHover"
}
```

Key keyboard shortcuts to memorize: - `Tab` — Accept suggestion - `Esc` — Dismiss suggestion - `Alt+]` / `Alt+[` — Cycle through alternative suggestions - `Ctrl+Enter` — Open Copilot completions panel (see multiple suggestions) -

Ctrl+I — Open inline chat (Copilot Chat) - Ctrl+Shift+I — Open Copilot Chat panel

Installing Claude Code

Claude Code runs in your terminal and is installed via npm.

Prerequisites

- Node.js 18 or later
- An Anthropic account with API access

Installation

```
## Install Claude Code globally
npm install -g @anthropic-ai/claude-code

## Verify the installation
claude --version

## Authenticate — just run `claude` and it will open a browser-based
## OAuth flow on first launch. There is no separate auth subcommand.
claude
```

Follow the browser-based authentication flow when prompted. Once authenticated, Claude Code stores your credentials securely.

First Run

Navigate to a project directory and run:

```
cd /path/to/your/project
claude
```

On first run in a new project, Claude Code will scan your codebase to build context. This takes a few seconds for small projects and up to a minute for large ones.

Configuration: The CLAUDE.md File

This is the single most impactful configuration you can do for Claude Code. Create a `CLAUDE.md` file in your project root:

```
## CLAUDE.md

### Project Overview
This is a SaaS application for inventory management built with:
- Frontend: React 18 + TypeScript + Tailwind CSS
- Backend: Node.js + Express + TypeScript
- Database: PostgreSQL with Prisma ORM
- Testing: Vitest (unit), Playwright (e2e)
- Deployment: Docker + AWS ECS

### Code Conventions
- Use functional components with hooks (no class components)
- All API responses follow the { data, error, meta } envelope pattern
- Error handling uses custom AppError class hierarchy
- Use named exports, not default exports
- Database queries go through repository classes, never raw SQL in routes

### Architecture
- /src/api/ — Express routes and middleware
- /src/services/ — Business logic (services call repositories)
- /src/repositories/ — Database access via Prisma
- /src/components/ — React components
- /src/hooks/ — Custom React hooks
- /src/utils/ — Shared utilities

### Testing Requirements
- All service functions must have unit tests
- All API endpoints must have integration tests
- Use factories for test data (see /tests/factories/)
- Never mock the database in integration tests — use test database

### Common Commands
- `npm run dev` — Start development server
- `npm run test` — Run all tests
- `npm run test:unit` — Run unit tests only
- `npm run lint` — Run ESLint
- `npm run db:migrate` — Run database migrations
```

This file is your project's instruction manual for AI. Every time Claude Code reads your project, it processes this file first. It turns generic AI assistance into project-aware AI assistance.

Tip: Other developers on your team benefit from this file too. Commit it to your repo. It doubles as human documentation.

VS Code Integration

Claude Code also has a VS Code extension that provides a panel inside the editor:

```
## Install the VS Code extension
code --install-extension anthropic.claude-code
```

This gives you Claude Code's agentic capabilities without leaving VS Code.

Setting Up a Chat Assistant

For architecture discussions, design exploration, and learning, you want a conversational AI on demand.

Option A: Claude (Web or Desktop App)

Sign up at claude.ai. The Pro plan ($20/month) gives you access to Claude Sonnet and Opus models with generous usage limits.

The Claude desktop app provides system-wide access via keyboard shortcut, so you can summon it from any context — reading documentation, reviewing designs, or browsing error logs.

Option B: ChatGPT (Web or Desktop App)

Sign up at chat.openai.com. The Plus plan ($20/month) gives you GPT-5.4 and canvas for code editing.

Option C: Gemini

Available at gemini.google.com. Best if you are deep in the Google ecosystem.

Recommendation: Pick one and stick with it. You build up conversation history and context over time. I use Claude for this role because it handles nuanced technical discussions well and writes clearly, but all three are competent.

Setting Up Cursor (Optional, Alternative to VS Code + Copilot)

If you prefer the all-in-one approach, Cursor replaces both VS Code and Copilot with a single AI-native editor.

Installation

Download from cursor.com. It imports your VS Code settings, extensions, and keybindings automatically.

Configuration

```
// Cursor settings (Settings → Cursor Settings)
{
  // Choose your preferred model for chat
  "cursor.ai.model": "claude-sonnet",

  // Enable codebase indexing for @codebase references
  "cursor.ai.indexCodebase": true,

  // Tab completion model
  "cursor.ai.autocomplete.model": "cursor-small"
}
```

Key shortcuts: - `Cmd+K` / `Ctrl+K` — Inline edit (select code, describe changes) - `Cmd+L` / `Ctrl+L` — Open chat panel - `Cmd+Shift+L` / `Ctrl+Shift+L` — Add selection to chat context - `Cmd+I` / `Ctrl+I` — Composer (multi-file agent mode)

Project-Level AI Configuration

Beyond tool-specific settings, project-level files improve AI assistance across all tools.

.cursorrules

If you use Cursor, create a `.cursorrules` file in your project root. This is Cursor's equivalent of `CLAUDE.md`:

```
You are helping develop an inventory management SaaS application.

Tech stack: React 18, TypeScript, Node.js, Express, PostgreSQL, Prisma.

Follow these conventions:
- Functional components with hooks
- Named exports only
- API responses use { data, error, meta } envelope
- Services never import from routes; routes never import from repositories
- All dates are stored as UTC, displayed in user's timezone
```

.github/copilot-instructions.md

GitHub Copilot supports repository-level instructions:

```
## Copilot Instructions

This is a TypeScript project using React and Express.

### Coding Standards
- Use strict TypeScript (no `any` types)
- Prefer `const` over `let`
- Use async/await, not raw Promises
- Error messages should be user-friendly, not technical

### Patterns
- React components: functional with hooks
- API calls: use the `apiClient` wrapper in `/src/utils/apiClient.ts`
- Form validation: use Zod schemas
```

.editorconfig

Not AI-specific, but AI tools respect `.editorconfig`, and it prevents arguments about formatting:

```
root = true

[*]
indent_style = space
indent_size = 2
end_of_line = lf
charset = utf-8
trim_trailing_whitespace = true
insert_final_newline = true

[*.py]
indent_size = 4

[*.go]
indent_style = tab
```

The Multi-Tool Workflow

How these tools work together in practice:

Scenario: Implementing a New Feature

1. **Think it through** (Chat assistant): Open Claude or ChatGPT. Describe the feature. Discuss architectural options. Get a rough plan.

```
"I need to add a bulk import feature for inventory items.
Users upload a CSV, we validate it, show a preview, and then
import the valid rows. What's the best approach for handling
files up to 10MB with real-time progress?"
```

2. **Implement it** (Claude Code or Cursor agent): Switch to your terminal or editor agent. Give it the plan.

```
claude "Implement the CSV bulk import feature for inventory items.
- Add a POST /api/imports/upload endpoint that accepts CSV files up to 10MB
- Parse and validate the CSV using Papa Parse
- Store import jobs in a new ImportJob table (id, status, filename,
  total_rows, valid_rows, error_rows, created_at)
- Return a job ID for polling progress
- Add a GET /api/imports/:id endpoint for checking import status
- Add a React component for the upload UI with progress display
- Write unit tests for the CSV parsing and validation logic"
```

3. **Refine inline** (Copilot): As you review and edit the generated code, Copilot helps with small adjustments, error handling, and filling in details.

4. **Verify** (Your brain): Read the code. Run the tests. Check for security issues. Confirm it matches your architecture.

This workflow takes an hour. Without AI, it takes a day. The key is using each tool for what it does best.

Environment Variables and Secrets

Quick but important: AI tools can see your environment variables, configuration files, and potentially your secrets. Be careful.

Do: - Use `.env` files and add them to `.gitignore` - Use environment variable references in code (not hardcoded values) - Keep AI tool API keys in your system keychain or credential store

Do not: - Paste API keys into AI chat conversations - Let AI tools commit `.env` files - Store secrets in `CLAUDE.md` or `.cursorrules`

```
## Good: reference environment variables
DATABASE_URL=${DATABASE_URL}

## Bad: hardcode secrets in AI-visible config
DATABASE_URL=postgresql://admin:s3cret@prod-db:5432/myapp
```

Terminal Multiplexing for AI Workflows

If you use Claude Code alongside your normal development terminal, a multiplexer makes the workflow smooth:

```
## tmux layout for AI development
## Pane 1 (left, large): Your editor (vim/nvim) or file browser
## Pane 2 (right top): Claude Code session
## Pane 3 (right bottom): Regular terminal for git, tests, etc.

## Quick tmux setup:
tmux new-session -s dev
## Ctrl+B % (split vertical)
## Ctrl+B " (split horizontal in right pane)
```

On Windows, Windows Terminal's split pane feature works similarly. On macOS, iTerm2's pane splitting is excellent.

Verifying Your Setup

Run through this checklist to confirm everything works:

1. **Copilot**: Open a code file, start typing a function signature. Do you see ghost text suggestions? Accept one with Tab.
2. **Claude Code**: In your terminal, navigate to a project and run `claude "What does this project do?"`. Does it give a coherent summary?
3. **Chat assistant**: Open your chosen chat tool. Ask it to explain a concept. Does it respond clearly?
4. **Project context**: Ask Claude Code to make a small change. Does the output follow your project's conventions (as specified in `CLAUDE.md`)?

If all four work, you are ready. If any fail, the most common issues are:

- **Authentication expired**: Re-run `claude` to trigger the OAuth flow again, or re-sign into Copilot
- **Missing Node.js**: Claude Code requires Node.js 18+
- **Extension conflicts**: Disable other AI extensions if Copilot is not showing suggestions
- **Network/firewall**: Corporate networks sometimes block AI tool APIs

Setup Done; Use Begins

Your environment is set: inline completion for typing, an agentic tool for complex tasks, a chat assistant for thinking. Chapter 4 starts with the most fundamental AI-assisted task — generating code that actually works. Not code that compiles. Not code that looks right. Code that does what you need, handles edge cases, and does not introduce bugs. That distinction is the difference between AI making you faster and AI making you more dangerous.

Part II: Core Workflows

"Debugging is twice as hard as writing the code in the first place."

— Brian Kernighan

Chapter 4: Code Generation That Actually Works

Getting Useful Code from AI, Not Just Code That Compiles

There is a huge difference between code that compiles and code that works. There is an even bigger difference between code that works and code that belongs in your codebase. AI can produce all three — and it is your job to know which one you got.

This chapter teaches you how to prompt AI for code generation that produces production-quality results, not demos. We cover the anatomy of effective prompts, iterative refinement, and the verification habits that separate AI-assisted developers from AI-dependent developers.

The Prompt Quality Spectrum

AI code generation quality is almost entirely determined by your prompt. Here is the spectrum:

Level 1: Vague (Bad)

```
Write a function to process orders
```

What you get: A generic function that takes "an order" and does... something. Maybe it calculates a total. Maybe it saves to a database. The AI guesses, and it guesses wrong.

Level 2: Specific (Okay)

```
Write a Python function that takes an order dictionary with items,
quantities, and prices, calculates the subtotal, applies tax, and
returns the total.
```

What you get: A function that handles the happy path. No error handling, no edge cases, probably uses `float` for money (which is wrong).

Level 3: Comprehensive (Good)

```
Write a Python function `calculate_order_total` that:
- Takes a list of OrderItem dataclasses (product_id: str,
  quantity: int, unit_price: Decimal)
- Calculates subtotal as sum of (quantity * unit_price)
- Applies a tax rate (passed as parameter, Decimal, e.g., Decimal("0.08"))
- Applies an optional discount (flat amount or percentage,
  passed as Optional[Discount] dataclass)
- Returns an OrderTotal dataclass with subtotal, tax_amount,
  discount_amount, and total
- Uses Decimal for all monetary calculations (never float)
- Raises ValueError if quantity is negative or unit_price is negative
- Include type hints and a docstring
```

What you get: Production-quality code that handles edge cases, uses correct types, and fits into a real system.

The Level 3 prompt took about 60 seconds to write. The Level 1 prompt took 5 seconds. But the Level 3 prompt saves you 30 minutes of editing, debugging, and rewriting. This is the math of AI-assisted development: invest seconds in the prompt, save minutes (or hours) in the output.

The Anatomy of an Effective Code Prompt

Every effective code generation prompt has five elements:

1. What (Function signature and purpose)

State what you want: the function name, its parameters, and what it returns.

```
Create a TypeScript function `mergeUserProfiles` that takes
two UserProfile objects and returns a merged UserProfile.
```

2. How (Behavior specification)

Describe the logic. Be specific about rules, precedence, and edge cases.

```
Merge rules:
- Non-null values from the second profile override the first
- Arrays (like `tags`) are merged and deduplicated
- Nested objects (like `preferences`) are deep-merged
- `createdAt` always takes the earlier timestamp
- `updatedAt` always takes the later timestamp
```

3. Types (Data structures)

Either provide the types or tell the AI where to find them.

```
interface UserProfile {
  id: string;
  email: string;
  name: string | null;
  tags: string[];
  preferences: Record<string, unknown>;
  createdAt: Date;
  updatedAt: Date;
}
```

4. Constraints (What NOT to do)

Negative constraints prevent common AI mistakes.

```
Constraints:
- Do not mutate either input object
- Do not use lodash or any external library
- Do not use `any` type
- Handle the case where either input could have undefined fields
```

5. Context (Where this lives)

Tell the AI about your project's patterns so the output fits in.

```
This function goes in src/utils/profile-utils.ts.
We use named exports. Other utilities in this directory
follow a similar pattern — see mergeConfigs in config-utils.ts.
```

Here is a complete prompt using all five elements:

```
Create a TypeScript function `mergeUserProfiles` in
src/utils/profile-utils.ts.

It takes two UserProfile objects (primary and secondary)
and returns a merged UserProfile.

Merge rules:
- Non-null values from secondary override primary
- Arrays (tags) are merged and deduplicated
- Nested objects (preferences) are deep-merged
- createdAt takes the earlier timestamp
```

```
- updatedAt takes the later timestamp
- id always comes from primary

Use the UserProfile interface from src/types/user.ts.
Do not mutate inputs. Do not use external libraries. No `any` type.
Use named export. Include JSDoc. Write 2-3 unit tests.
```

This prompt produces code you can drop into your project with minimal editing.

Language-Specific Prompting Strategies

Different languages have different gotchas. Adjust your prompts accordingly.

Python

AI loves writing Python, and Python has common pitfalls to guard against:

```
Write a Python class `RateLimiter` that implements a
token bucket algorithm.

Requirements:
- Use dataclasses, not regular classes with __init__
- Type hints on all methods (use `from __future__ import annotations`)
- Use `time.monotonic()` not `time.time()` for timing
- Thread-safe using threading.Lock
- Do NOT use mutable default arguments
- Compatible with Python 3.11+
```

The "do NOT use mutable default arguments" line prevents a classic Python bug that AI regularly introduces:

```python
## BAD — AI generates this constantly
def process_items(items: list[str], seen: set = set()):
    # `seen` is shared across ALL calls! Bug!
    ...

## GOOD — what you actually want
def process_items(items: list[str], seen: set | None = None):
    if seen is None:
        seen = set()
    ...
```

TypeScript

TypeScript prompts should emphasize type safety:

```
Write a TypeScript function `parseApiResponse<T>` that:
- Takes a raw Response object from fetch
- Validates the response status (throws on non-2xx)
- Parses the JSON body
- Validates the body against a Zod schema (passed as parameter)
- Returns the validated, typed data

Use generics properly. The return type should be inferred
from the Zod schema. No `any`, no `as` type assertions,
no `@ts-ignore`.
```

That last line is critical. AI-generated TypeScript is riddled with `as any` casts and `@ts-ignore` comments that silently defeat type safety.

```
// What AI often generates (BAD)
const data = JSON.parse(body) as UserProfile;  // No runtime validation!

// What you want (GOOD)
const data = userProfileSchema.parse(JSON.parse(body));  // Validates at runtime
```

Go

Go prompts should emphasize error handling and idiomatic patterns:

```
Write a Go function `FetchUserByEmail` that:
- Queries a PostgreSQL database using pgx (not database/sql)
- Returns (*User, error)
- Wraps database errors with fmt.Errorf and %w verb
- Returns a custom ErrUserNotFound for missing users
  (not a generic error)
- Uses context.Context as first parameter
- Handles pgx.ErrNoRows explicitly
- Do not use panic for error handling
```

```
// What AI often generates (BAD)
func FetchUserByEmail(email string) User {
    var user User
    err := db.QueryRow("SELECT * FROM users WHERE email = ?", email).Scan(&user)
    if err != nil {
        panic(err)  // Never do this
    }
```

```
    return user
}

// What you want (GOOD)
func FetchUserByEmail(ctx context.Context, db *pgx.Conn, email string) (*User, e
rror) {
    var user User
    err := db.QueryRow(ctx,
        "SELECT id, email, name, created_at FROM users WHERE email = $1",
        email,
    ).Scan(&user.ID, &user.Email, &user.Name, &user.CreatedAt)
    if err != nil {
        if errors.Is(err, pgx.ErrNoRows) {
            return nil, ErrUserNotFound
        }
        return nil, fmt.Errorf("fetching user by email %q: %w", email, err)
    }
    return &user, nil
}
```

Iterative Refinement

No prompt produces perfect code on the first try every time. Iterative refinement is not a failure — it is the process.

The Three-Pass Approach

Pass 1: Generate. Write your best prompt. Get the initial output.

Pass 2: Critique. Read the output critically. Identify issues. Ask the AI to fix them:

```
This looks good, but three issues:
1. The error messages expose internal database column names —
   use generic user-facing messages instead
2. The `processItems` function doesn't handle empty arrays
3. You're using `any` in the error handler — use `unknown`
   and narrow the type
```

Pass 3: Integrate. Once the code is solid, adapt it to your codebase. That might mean renaming variables to match your conventions, dropping it in the right file, or adjusting imports.

When to Iterate vs. When to Rewrite

Iterate when: - The structure is right but details are wrong - You need to add edge case handling - The code needs to match your conventions

Rewrite (start a new prompt) when: - The fundamental approach is wrong - The AI chose the wrong algorithm or pattern - You have learned enough from the first attempt to write a much better prompt

Common AI Code Generation Failures

Knowing how AI fails helps you catch problems before they ship.

Hallucinated APIs

AI invents function signatures, library methods, and entire packages that do not exist. Especially common with newer libraries and less popular languages.

```
## AI might generate this (WRONG)
from fastapi.security import OAuth2BearerToken  # This class doesn't exist

## The actual API
from fastapi.security import OAuth2PasswordBearer
```

Prevention: Always verify imports. If you do not recognize a class or function name, check the library documentation before trusting it.

Outdated Patterns

AI training data has a cutoff. Code patterns change. What was idiomatic in 2023 might be deprecated in 2026.

```
// AI might suggest (OUTDATED)
const router = require('express').Router();  // CommonJS

// Current best practice
import { Router } from 'express';  // ESM
```

Prevention: Specify the version or era in your prompt: “Use React 18 patterns” or “Use Express 5 with ESM imports.”

Subtly Wrong Logic

The most dangerous failure mode. The code looks right, passes basic tests, fails on edge cases.

```
## AI-generated discount calculation (WRONG)
def apply_discount(price: Decimal, discount_percent: int) -> Decimal:
    return price * (1 - discount_percent / 100)
    # Bug: discount_percent / 100 produces a float, and multiplying
    # a Decimal by a float raises TypeError in Python 3. Even if you
    # work around it, mixing Decimal with float arithmetic silently
    # introduces floating-point imprecision — exactly what Decimal
    # exists to avoid.
    # Also: doesn't handle discount > 100% or negative discounts

## Correct version — keep everything in Decimal
def apply_discount(price: Decimal, discount_percent: Decimal) -> Decimal:
    if discount_percent < 0 or discount_percent > 100:
        raise ValueError(f"Invalid discount: {discount_percent}%")
    return price * (Decimal("1") - discount_percent / Decimal("100"))
```

Prevention: Write tests before or alongside the implementation. Tell the AI to generate tests too, then add your own edge case tests.

Security Holes

AI-generated code routinely has security vulnerabilities. Serious enough that Chapter 17 is dedicated to it, but the highlights:

- SQL injection (string concatenation instead of parameterized queries)
- Missing input validation
- Hardcoded secrets in examples that make it to production
- Insecure defaults (HTTP instead of HTTPS, permissive CORS)
- Missing authentication checks

```
## AI often generates (VULNERABLE)
@app.route("/users/<user_id>")
def get_user(user_id):
    query = f"SELECT * FROM users WHERE id = '{user_id}'"  # SQL injection!

## What you need (SAFE)
@app.route("/users/<int:user_id>")
@require_auth
```

```python
def get_user(user_id: int):
    user = db.session.get(User, user_id)  # Parameterized by ORM
    if user is None:
        abort(404)
    if user.id != current_user.id and not current_user.is_admin:
        abort(403)  # Authorization check!
    return jsonify(user.to_dict())
```

The Verification Checklist

After every AI code generation, run through this checklist:

1. **Read every line.** Do you understand what each line does?
2. **Check imports.** Do all imported modules and functions exist?
3. **Verify types.** Are types correct? Any `any`, `as`, or type assertions hiding?
4. **Test the happy path.** Does basic input produce correct output?
5. **Test edge cases.** Empty inputs, null values, boundary conditions, very large inputs.
6. **Check error handling.** What happens when things go wrong? Are errors caught and handled?
7. **Review security.** Input validation? SQL injection? Authentication? Authorization?
8. **Confirm it fits.** Does it follow your project's patterns and conventions?

This takes 5–10 minutes. It saves hours of debugging later.

Practical Example: Full Feature Generation

Let's walk through a real feature from start to finish.

The need: A rate limiter middleware for an Express API.

The prompt:

```
Write an Express middleware for rate limiting with these requirements:

- Token bucket algorithm
- Configurable: max tokens, refill rate, refill interval
- Per-client limiting using IP address (from req.ip)
  or API key (from x-api-key header, preferred if present)
- In-memory store (Map) with automatic cleanup of expired entries
```

```
- Returns 429 Too Many Requests with Retry-After header when limited
- Returns X-RateLimit-Remaining and X-RateLimit-Limit headers on all responses
- TypeScript, strict types, no `any`
- Export a factory function `createRateLimiter(options)` that returns the middle
ware
- Include the RateLimiterOptions interface
- Include unit tests using vitest

This goes in src/middleware/rate-limiter.ts. We use named exports.
Our other middleware follows the same factory pattern — see
src/middleware/auth.ts for reference.
```

The AI will generate something close to production-ready. You will need to:

1. Verify the token bucket implementation is correct (refill logic is tricky)
2. Ensure the cleanup interval does not leak (clearInterval on server shutdown)
3. Add the middleware to your route configuration
4. Run the tests and add any missing edge cases

This is the workflow. Specify precisely, generate quickly, verify thoroughly. That is how AI code generation works — not as a magic wand, but as a power tool that requires a skilled operator.

Generation is only half the story. Chapter 5 covers the other half — using AI as a code reviewer, what it catches that humans miss, and what it misses that you need to catch yourself.

Chapter 5: Code Review with AI

Using AI as a Code Reviewer — What It Catches and What It Misses

Code review is one of the highest-leverage activities in software development. It catches bugs before they ship, maintains code quality, spreads knowledge across the team, and prevents architectural drift. It is also slow, tedious, and frequently inconsistent.

AI does not replace human code review. But it makes human code review dramatically better by handling the mechanical parts — style consistency, common bug patterns, missing edge cases — so that human reviewers can focus on the parts that require human judgment: architecture, design intent, and business logic correctness.

AI as Your First Reviewer

The best workflow treats AI as your first reviewer and a human as your second. Before you open a pull request, before you ask a colleague to spend 30 minutes on your code, run it through AI review. This catches the embarrassing stuff — the missing null check, the inconsistent naming, the function doing too many things — before a human ever sees it.

Using Claude Code for Pre-PR Review

Claude Code can review changes across your entire working tree:

```
## Review all staged changes
claude "Review my staged changes. Look for bugs, security issues,
  performance problems, and code style violations. Be specific
  about line numbers and what's wrong."

## Review a specific file
claude "Review src/services/payment-service.ts. Focus on error
  handling and security. This processes real credit card transactions."

## Review against your project standards
```

```
claude "Review my changes against our project conventions in CLAUDE.md.
  Flag any violations."
```

Using Copilot Chat for In-Editor Review

In VS Code, you can select code and ask Copilot to review it:

1. Select the code you want reviewed
2. Press `Ctrl+I` to open inline chat
3. Type: `/review` or "Review this code for bugs and improvements"

Copilot Chat will analyze the selected code and provide inline feedback.

Using ChatGPT or Claude Web for Detailed Review

For complex logic that needs thorough analysis, paste the code into a chat assistant with context.

```
I'm implementing a distributed lock using Redis. Review this code
for correctness, race conditions, and failure modes. I'm particularly
worried about the lock release logic — is it safe if the process
crashes between acquiring the lock and setting the expiry?

[paste code]
```

The conversational format lets you ask follow-up questions, valuable for understanding why something is a problem, not just that it is.

What AI Catches Well

AI code review excels at pattern-matching — finding instances of known bad patterns. Here are the categories where AI consistently outperforms casual human review:

Null/Undefined Handling

AI is relentless about missing null checks. Humans glance at code and assume the happy path. AI examines every path.

```
// AI will flag this immediately
async function getOrderTotal(orderId: string): Promise<number> {
  const order = await db.orders.findUnique({ where: { id: orderId } });
  return order.items.reduce((sum, item) => sum + item.price, 0);
```

```
  // Bug: order could be null if orderId doesn't exist
  // Bug: order.items could be an empty array (returns 0, but is that correct?)
}
```

Resource Leaks

AI catches unclosed connections, file handles, and event listeners:

```
## AI will flag the missing connection cleanup
def fetch_user_data(user_id: int) -> dict:
    conn = psycopg2.connect(DATABASE_URL)
    cursor = conn.cursor()
    cursor.execute("SELECT * FROM users WHERE id = %s", (user_id,))
    result = cursor.fetchone()
    return dict(result)
    # Bug: connection and cursor are never closed
    # If an exception occurs, they leak
```

AI's suggestion:

```
def fetch_user_data(user_id: int) -> dict:
    with psycopg2.connect(DATABASE_URL) as conn:
        with conn.cursor(cursor_factory=RealDictCursor) as cursor:
            cursor.execute("SELECT * FROM users WHERE id = %s", (user_id,))
            result = cursor.fetchone()
            if result is None:
                raise UserNotFoundError(f"User {user_id} not found")
            return dict(result)
```

Common Security Vulnerabilities

AI has seen thousands of examples of SQL injection, XSS, and other vulnerability patterns. It catches them more reliably than tired human reviewers do:

```
// AI flags this as SQL injection
app.get('/search', (req, res) => {
  const query = `SELECT * FROM products WHERE name LIKE '%${req.query.q}%'`;
  db.query(query, (err, results) => {
    res.json(results);
  });
});

// AI flags this as XSS
app.get('/profile', (req, res) => {
```

```
  res.send(`<h1>Welcome, ${req.query.name}</h1>`);
});
```

Error Swallowing

AI catches empty catch blocks and errors that are logged but not handled:

```
// AI will flag this pattern
data, err := json.Marshal(response)
if err != nil {
    log.Println("marshal error:", err)
    // Bug: function continues with nil/zero data
    // The error is logged but not returned or handled
}
w.Write(data)  // Writes nil if marshal failed
```

Inconsistent Error Handling

AI notices when different parts of your code handle the same category of errors differently:

```
// In route A, you throw HttpError
if (!user) throw new HttpError(404, "User not found");

// In route B, you return an error response directly
if (!product) return res.status(404).json({ error: "Not found" });

// In route C, you call next() with an error
if (!order) return next(new NotFoundError("Order"));

// AI will flag all three as inconsistent and ask which pattern
// you want to standardize on
```

Performance Issues

AI catches common performance anti-patterns:

```
## AI flags the N+1 query problem
async def get_users_with_posts(db: AsyncSession) -> list[UserResponse]:
    users = await db.execute(select(User))
    result = []
    for user in users.scalars():
        # This executes a separate query for EACH user
        posts = await db.execute(
            select(Post).where(Post.user_id == user.id)
```

```
        )
        result.append(UserResponse(
            user=user,
            posts=list(posts.scalars())
        ))
    return result
```

What AI Misses

The critical section. Knowing what AI does not catch well is what prevents you from over-relying on it.

Business Logic Correctness

AI does not know your business rules. It cannot tell you that your discount calculation is wrong because premium customers get a different rate than standard customers. It sees syntactically valid code and calls it fine.

```
def calculate_subscription_price(
    base_price: Decimal,
    plan: str,
    months: int
) -> Decimal:
    if months >= 12:
        discount = Decimal("0.20")  # 20% annual discount
    elif months >= 6:
        discount = Decimal("0.10")  # 10% semi-annual discount
    else:
        discount = Decimal("0")

    return base_price * months * (1 - discount)
```

AI will say this looks correct. But your business team wants annual billing calculated at 10 months' price (not 12 months minus 20%), and enterprise plans get a different discount schedule entirely. Only a human who knows the business rules catches this.

Architectural Violations

AI might not know that your team decided services should never import from controllers, or that database access must go through the repository layer. Unless you have documented these rules in your AI configuration files (CLAUDE.md, .cursorrules), AI will not flag violations.

```
// This might be a serious architectural violation
// but AI won't know unless you tell it about your architecture
import { prisma } from '../lib/prisma'; // Direct DB access in a controller!

export class UserController {
  async getUser(req: Request, res: Response) {
    const user = await prisma.user.findUnique({ // Should go through UserServic
e
      where: { id: req.params.id }
    });
  }
}
```

Fix: Document your architectural rules in CLAUDE.md and .cursorrules. Then AI can enforce them.

Concurrency Issues

AI struggles with race conditions, deadlocks, and other concurrency bugs because they depend on execution timing, not code structure.

```
// This has a race condition, but AI often misses it
var cache = make(map[string]interface{})

func GetCachedValue(key string) interface{} {
    if val, ok := cache[key]; ok {
        return val
    }
    val := expensiveLookup(key)
    cache[key] = val  // Race condition: concurrent writes to map
    return val
}
```

AI sometimes catches obvious concurrency bugs (unsynchronized map access in Go is a known pattern), but it often misses subtler issues like TOCTOU (time-of-check-to-time-of-use) races and lock ordering problems.

Design Intent

AI can tell you whether code does what it does correctly. It cannot tell you whether it should be doing that thing at all.

A function might be perfectly implemented but wrong — solving the wrong problem, building a feature that should not exist, or implementing a workaround for a bug that should be fixed at the root cause.

Integration Context

AI sees individual files or small groups of files. It often misses how changes affect other parts of the system — downstream consumers, database triggers, background jobs that depend on certain data formats, external APIs that expect specific request shapes.

The Hybrid Review Workflow

The workflow I recommend for teams:

Step 1: Author Self-Review with AI

Before opening a PR, the author runs AI review on their own changes:

```
## Stage your changes, then review.
## Use specific file paths rather than `git add -A`, which can
## accidentally stage secrets, build artifacts, or other files
## you do not intend to commit. If you do use `git add -A`,
## make sure your .gitignore is properly configured first.
git add <files you intend to commit>
claude "Review my staged changes. Check for:
1. Bugs and edge cases
2. Security vulnerabilities
3. Performance issues
4. Violations of our project conventions (see CLAUDE.md)
5. Missing tests
Be specific about file names and line numbers."
```

Fix everything AI finds. This takes 10–15 minutes and eliminates 60–70% of the issues a human reviewer would catch.

Step 2: AI Review on the PR

Many teams set up automated AI review on pull requests. GitHub Copilot for Business includes PR review features. Several third-party tools (CodeRabbit, Sourcery, Ellipsis) provide automated PR review.

These tools comment directly on your PR with specific feedback.

```
## Automated review comment on a PR diff

> src/services/order-service.ts, line 47:
> The `calculateTotal` function doesn't handle the case where
> `items` is an empty array. This will return `NaN` because
> `reduce` on an empty array without an initial value throws.
>
> Suggestion: Add an initial value to reduce, or check for
> empty arrays first.
```

Step 3: Human Review

The human reviewer now sees a PR already cleaned up. They can focus on:

- Does this approach make sense architecturally?
- Does it match the requirements?
- Are there better alternatives?
- Is this the right level of abstraction?
- Are there implications for other parts of the system?

A dramatically better use of a senior developer's time than catching missing null checks.

Prompting AI for Better Reviews

The quality of AI review, like all AI work, depends on the prompt. Patterns that produce useful reviews:

Role-Based Review

```
Review this code as a senior security engineer. Focus exclusively
on security vulnerabilities, input validation, authentication,
authorization, and data exposure risks. Ignore style issues.
```

Checklist-Based Review

```
Review this pull request against this checklist:
- [ ] All public functions have error handling
- [ ] No raw SQL queries (use ORM)
- [ ] All user inputs are validated before use
- [ ] No console.log statements (use logger)
```

```
- [ ] New database queries have appropriate indexes
- [ ] API responses don't leak internal implementation details
```

Comparative Review

```
Here are two implementations of the same function. Which is better
and why? Consider performance, readability, and maintainability.

Implementation A:
[code]

Implementation B:
[code]
```

Devil's Advocate Review

```
Try to break this code. What inputs would cause it to fail?
What assumptions does it make that might be wrong? What happens
under load? What happens if the database is slow? What happens
if the external API returns unexpected data?
```

Building an AI Review Culture

If you are introducing AI review to a team, the practices that work:

Make it the norm, not the exception. AI review should be a standard step in the PR process, not something individuals optionally do.

Do not let AI review replace human review. AI review reduces the burden on human reviewers. It does not eliminate the need for them.

Track what AI catches. Keep a record of the issues AI finds. Over time, this surfaces patterns in your team's common mistakes, which you can address with training, linting rules, or project configuration changes.

Keep your AI configuration files updated. As your project evolves, update CLAUDE.md and .cursorrules with new conventions. AI review is only as good as the context you give it.

Review the AI's review. Occasionally, AI review throws false positives — it flags something as a bug when the code is correct, or suggests a "fix" that introduces a new problem. Apply judgment to AI feedback, just as you do with human feedback.

The ROI of AI-Assisted Code Review

Some rough numbers. For a team of 8 developers, each opening 3–4 PRs per week:

- **Without AI review:** Each PR takes 20–30 minutes of reviewer time. That is 480–960 minutes (8–16 hours) of review time per week across the team.
- **With AI review:** AI handles the mechanical review in 2–3 minutes. Human review drops to 10–15 minutes because the easy stuff is already caught. Total: 240–480 minutes (4–8 hours) per week.

A 50% reduction in review time, with higher-quality reviews because humans focus on what matters.

More importantly, the feedback loop is faster. Instead of waiting hours for a human reviewer to find a missing null check, the author catches it in 2 minutes with AI review before the PR is even opened. Faster feedback means faster iteration, which means faster shipping.

Chapter 6: Debugging with AI

Feeding Errors, Logs, and Context to AI for Faster Diagnosis

Every developer has stared at an error message for 20 minutes, Googled it, read three Stack Overflow threads that are almost but not quite the same problem, tried the top answer, broken something else, and finally realized the issue was a missing comma on line 47.

AI does not eliminate debugging. But it sharply compresses the time between "what is this error" and "oh, that's the problem." AI has seen your error before — thousands of times, in thousands of contexts. Its pattern recognition for common bugs is genuinely superhuman.

This chapter teaches you how to feed bugs to AI effectively, how to use it across different categories of debugging, and when AI debugging hits its limits.

The Debugging Prompt Formula

Effective debugging prompts follow a consistent structure:

```
I'm getting this error:
[exact error message / stack trace]

In this code:
[relevant code, with surrounding context]

Expected behavior:
[what should happen]

Actual behavior:
[what actually happens]

What I've tried:
[any debugging steps you've already taken]

Environment:
[language version, framework version, OS, etc.]
```

Every element matters. Here's why, with a real example.

Bad Debugging Prompt

```
My React app is crashing. Help.
```

AI's response will be generic and useless — a list of common React errors, most of which won't apply.

Good Debugging Prompt

```
I'm getting this error in my React 18 app with TypeScript:

Error: Rendered more hooks than during the previous render.
    at renderWithHooks (react-dom.development.js:16325)
    at updateFunctionComponent (react-dom.development.js:19588)
    at beginWork (react-dom.development.js:21601)

The error occurs in this component:

function UserProfile({ userId }: { userId: string }) {
  const { data: user } = useQuery({
    queryKey: ['user', userId],
    queryFn: fetchUser,
  });

  if (!user) return <Loading />;

  const { data: orders } = useQuery({
    queryKey: ['orders', userId],
    queryFn: () => fetchOrders(userId),
  });

  return <div>{user.name}: {orders?.length} orders</div>;
}

Expected: Component renders user profile with order count
Actual: Crashes with the hooks error on second render

Using React 18.2, TanStack Query v5, TypeScript 5.3
```

AI spots the problem instantly: the second `useQuery` hook is called conditionally — after the early return — which violates React's rules of hooks. Every hook must be called on every render, in the same order.

```
// Fixed version — both hooks called unconditionally
function UserProfile({ userId }: { userId: string }) {
  const { data: user } = useQuery({
    queryKey: ['user', userId],
    queryFn: fetchUser,
  });
  const { data: orders } = useQuery({
    queryKey: ['orders', userId],
    queryFn: () => fetchOrders(userId),
    enabled: !!user, // Conditionally fetch, but hook always called
  });

  if (!user) return <Loading />;

  return <div>{user.name}: {orders?.length ?? 0} orders</div>;
}
```

The difference between the bad and good prompts: 30 seconds of typing, 20 minutes less debugging.

Debugging Categories and AI Effectiveness

Different categories of bugs yield different results. Knowing which is which helps you calibrate expectations.

Syntax and Type Errors: AI Rating — Excellent

AI demolishes syntax errors. This is its strongest debugging domain.

```
## TypeScript error
src/utils/parser.ts:23:5 - error TS2345: Argument of type 'string' is
not assignable to parameter of type 'number'.
```

Feed the error and the relevant code to AI. It spots the type mismatch and suggests the correct fix in seconds — including the gnarly generic type errors that are notoriously hard to read:

```
Type 'Observable<HttpEvent<T>>' is not assignable to type
'Observable<HttpEvent<HttpResponse<T>>>'.
  Type 'HttpEvent<T>' is not assignable to type
  'HttpEvent<HttpResponse<T>>'.
```

AI decodes these nested type errors better than most developers can.

Runtime Exceptions: AI Rating — Very Good

Stack traces with clear error messages are AI's bread and butter:

```
Traceback (most recent call last):
  File "app/services/order_service.py", line 45, in process_order
    total = sum(item.price * item.quantity for item in order.items)
  File "app/services/order_service.py", line 45, in <genexpr>
    total = sum(item.price * item.quantity for item in order.items)
TypeError: unsupported operand type(s) for *: 'str' and 'int'
```

AI sees instantly that `item.price` is a string — probably from JSON parsing — and needs conversion to a numeric type. It will also note that monetary calculations should use `Decimal`, not `float`.

Logic Bugs: AI Rating — Good (With Context)

Logic bugs — code that runs without errors but produces wrong results — are harder because there's no error message to key off. The quality of AI's diagnosis depends entirely on the context you provide.

```
## "This function is supposed to check if a user can access a resource,
## but it's returning True for users who should be denied."

def can_access(user: User, resource: Resource) -> bool:
    if user.role == "admin":
        return True
    if resource.owner_id == user.id:
        return True
    if resource.visibility == "public":
        return True
    if user.role in resource.allowed_roles:
        return True
    return False
```

AI might ask: "What roles does the user have? What are the allowed_roles on the resource?" Then it might spot the issue — perhaps `user.role` is a single string but `resource.allowed_roles` is a list, and the `in` check matches a substring instead of an exact role. Or perhaps `resource.visibility` defaults to `"public"` when it should default to `"private"`.

Concurrency Bugs: AI Rating — Fair

Race conditions, deadlocks, and timing-dependent bugs are hard for AI because they depend on execution order, not code structure. AI can suggest common concurrency patterns and spot obvious issues — unsynchronized access to shared state, for instance — but it cannot simulate concurrent execution.

```
// "This occasionally panics with 'concurrent map writes'"
var cache sync.Map  // AI will suggest using sync.Map if you're using a regular map

// But for subtler issues like this:
func Transfer(from, to *Account, amount int) {
    from.mu.Lock()
    to.mu.Lock()  // Potential deadlock if another goroutine locks in opposite order
    defer from.mu.Unlock()
    defer to.mu.Unlock()
    // ...
}
```

AI will recognize the classic lock-ordering deadlock and suggest always locking in a consistent order — for example, by account ID.

Environment-Specific Bugs: AI Rating — Variable

"It works on my machine" bugs depend on environment details AI cannot observe. Give it as much environmental context as possible:

```
This code works in development (macOS, Node 20, PostgreSQL 15 via Docker)
but fails in CI (Ubuntu 22.04, Node 20, PostgreSQL 14 via GitHub Actions).

Error: relation "users" does not exist

The migration runs successfully — I can see it in the CI logs.
But the test that runs immediately after gets this error.
```

AI might suggest the migration is running against a different database than the tests, or that there's a schema search path issue between PostgreSQL 14 and 15, or that CI is running tests before migrations complete. The diagnosis is educated guessing based on common CI patterns.

Performance Bugs: AI Rating — Good for Obvious, Poor for Subtle

AI catches obvious performance issues — N+1 queries, missing indexes, unnecessary re-renders — but struggles with subtle performance problems that require profiling data.

```
## Feed AI your slow query with the EXPLAIN output
"This query takes 4 seconds in production but runs instantly
in development. Here's the EXPLAIN ANALYZE output:"

[paste explain output]

"The users table has 2 million rows in production and 100
rows in development."
```

AI will analyze the execution plan and likely flag a missing index or a query that scans the entire table. Highly practical.

Tool-Specific Debugging Workflows

Claude Code: Codebase-Aware Debugging

Claude Code's ability to read your entire codebase makes it uniquely powerful for debugging:

```
## Claude Code can follow the call chain
claude "I'm getting a 500 error on POST /api/orders. The error
  message is 'Cannot read properties of undefined (reading map)'.
  Trace through the code from the route handler to find where
  the undefined value originates."

## Claude reads the route, the service, the repository,
## and any middleware in the chain to find the root cause.
```

This is dramatically better than pasting individual files into a chat. Claude Code sees the entire execution path.

Copilot: Inline Error Resolution

When you see a red squiggle in your editor, hover over it and click "Quick Fix" — Copilot often nails the fix. For thornier errors, use the inline chat:

1. Click on the error line

2. Press Ctrl+I
3. Type: "Fix this error: [paste the error message]"

Chat Assistants: Deep Debugging Conversations

For bugs that need back-and-forth investigation, a chat assistant is ideal:

```
Me: This webhook handler sometimes processes the same event
    twice. Here's the code. [paste]

AI: I see the issue – there's no idempotency check. If the
    webhook fires twice (which happens under network issues),
    your handler processes both.

Me: How do I add idempotency? We use PostgreSQL.

AI: [provides implementation with a unique constraint on event_id]

Me: What about webhook events that arrive out of order?

AI: [provides implementation with event sequence checking]
```

The conversation surfaces deeper issues that a single prompt would miss.

Log Analysis with AI

Production logs are a goldmine of debugging information, but reading them manually is painful. AI excels at log analysis.

Pattern: Feed Relevant Logs

```
Here are the last 50 lines of logs before the crash.
The application is a Python FastAPI service processing
payment webhooks. It crashed at 14:23:07 UTC.

[2026-03-15 14:22:55] INFO: Webhook received: evt_abc123
[2026-03-15 14:22:55] INFO: Processing payment.succeeded
[2026-03-15 14:22:56] INFO: Updating order ord_xyz789
[2026-03-15 14:23:01] WARNING: Database connection pool exhausted
[2026-03-15 14:23:01] INFO: Webhook received: evt_def456
[2026-03-15 14:23:03] WARNING: Retry attempt 1 for order update
[2026-03-15 14:23:05] WARNING: Database connection pool exhausted
[2026-03-15 14:23:07] ERROR: Connection timeout after 30s
[2026-03-15 14:23:07] CRITICAL: Unhandled exception in webhook handler
```

```
What happened and how do I fix it?
```

AI will diagnose this as connection pool exhaustion — likely from long-running transactions holding connections while new webhooks keep arriving. It will suggest pool tuning, query timeouts, and webhook queuing.

Pattern: Comparing Working vs. Broken

```
Here's a successful request's log entries:
[logs from working request]

Here's a failing request's log entries:
[logs from failing request]

What's different and what's causing the failure?
```

This differential analysis is something AI does extremely well. It will pinpoint the divergence and suggest the cause.

The Debugging Decision Tree

Here is when to use each AI tool for debugging:

```
Is the error message clear and specific?
├── Yes → Feed it to Copilot inline or Claude Code
│         (resolution in < 2 minutes)
│
├── The error is clear but the cause is in another file
│   → Use Claude Code (it can follow cross-file dependencies)
│
├── No clear error, just wrong behavior
│   ├── Can you reproduce it with a test?
│   │   ├── Yes → Write the failing test, then ask AI
│   │   │         "why does this test fail?" with the code
│   │   └── No → Describe the behavior gap to a chat assistant
│   │            with as much context as possible
│   │
│   └── Is it a performance issue?
│       → Share profiling data / EXPLAIN output with AI
│
└── Intermittent / timing-dependent?
    → Describe the pattern (when it happens, when it doesn't)
```

```
to a chat assistant. AI might identify the trigger
even if it can't simulate the concurrency.
```

When AI Debugging Fails

Be honest with yourself about when AI isn't helping:

You've been going back and forth for 20 minutes. If AI hasn't identified the issue after three exchanges, it probably doesn't have enough context or the bug is outside its pattern recognition. Time to debug manually.

AI suggests a fix that doesn't work, then suggests another that doesn't work. This is the AI equivalent of trial and error. You can do trial and error faster yourself, because you can actually run the code and observe the results.

The bug is in the interaction between your code and a third-party system. AI may not know the third-party system's undocumented behaviors, version-specific bugs, or configuration quirks. Read the third-party docs or file an issue.

The bug requires understanding state over time. AI sees static code. Bugs that depend on accumulated state, session history, or complex user flows require you to trace state manually. Debugger breakpoints and logging are your friends here.

Building Debugging Prompts as Templates

Create template prompts for common debugging scenarios and save them:

```
## Save as a shell alias or snippet
alias debug-api='echo "API endpoint:
HTTP method:
Request body/params:
Expected response:
Actual response:
Status code:
Error message:
Relevant middleware:
Recent changes to this endpoint: "'
```

When you hit a bug, fill in the template and paste it into your AI tool. You'll provide consistent, comprehensive context every time.

From Today's Bugs to Yesterday's Debt

Debugging fixes today's bugs. But what about the accumulated weight of yesterday's bugs, shortcuts, and compromises? Chapter 7 covers using AI for refactoring and technical debt — taking code that mostly works and turning it into code that works well, reads clearly, and won't collapse under the next feature request.

Chapter 7: Refactoring and Technical Debt

Using AI to Modernize and Clean Up Codebases

Technical debt is the silent tax on every feature you build. It slows you down incrementally, so gradually that you don't notice — until one day a simple feature takes two weeks because the code it touches is a fragile, tangled mess.

AI is the best refactoring partner most developers have ever had. It's patient (it will refactor 200 files without complaining), knowledgeable (it knows the patterns you should be using), and fast (it can restructure a module in minutes). It's also dangerous — automated refactoring without understanding can break things in subtle ways.

This chapter teaches you how to use AI to systematically reduce technical debt, modernize legacy patterns, and refactor safely.

The AI Refactoring Advantage

Refactoring has always had a cost-benefit problem. The benefits are real — cleaner code, fewer bugs, faster development — but the cost is developer time spent on changes that don't add features. Managers see refactoring as time not spent building things.

AI changes the calculation. A refactor that would take two days takes two hours with AI assistance. The cost drops so dramatically that the ROI becomes obvious.

Here's where AI shines in refactoring:

Mechanical Transformations

Renaming a concept across a codebase. Changing function signatures. Moving code between files. Updating import paths. These are tedious, error-prone, and perfect for AI.

```
## Claude Code handles codebase-wide renames
claude "Rename the concept 'Customer' to 'Client' across the
  entire codebase. Update class names, variable names, file names,
  database references, API endpoints, and documentation.
  Preserve the git history by doing this as a single coherent change."
```

Pattern Modernization

Updating code to use modern language features:

```
claude "Refactor all callback-based async code in src/services/
  to use async/await. Preserve error handling behavior.
  Do not change any function's external interface."
```

Before:

```
function getUser(id, callback) {
  db.query('SELECT * FROM users WHERE id = ?', [id], (err, rows) => {
    if (err) {
      callback(err, null);
      return;
    }
    if (rows.length === 0) {
      callback(new NotFoundError('User not found'), null);
      return;
    }
    callback(null, rows[0]);
  });
}
```

After:

```
async function getUser(id) {
  const rows = await db.query('SELECT * FROM users WHERE id = ?', [id]);
  if (rows.length === 0) {
    throw new NotFoundError('User not found');
  }
  return rows[0];
}
```

Extract and Decompose

Breaking large functions into smaller, well-named functions:

```
This function is 200 lines long and does too many things.
Decompose it into smaller functions. Each function should:
```

```
- Do one thing
- Have a clear, descriptive name
- Be independently testable
- Not exceed 30 lines

Preserve the original function as a composed version that calls
the extracted functions, so the external interface doesn't change.
```

Dead Code Removal

AI can identify code that is never called:

```
claude "Analyze the codebase and identify functions, classes,
  and modules that appear to be unused (no imports, no calls,
  no references). List them with your confidence level.
  Do NOT delete anything yet - just report."
```

The analysis isn't perfect — AI may miss dynamic imports, reflection-based calls, or code referenced only from configuration files — but it gives you a strong starting list for cleanup.

Safe Refactoring with AI

The golden rule of refactoring: behavior must not change. Only the structure changes. AI sometimes breaks this rule, so you need safeguards.

Step 1: Ensure Test Coverage Before Refactoring

Never refactor code that doesn't have tests. If it doesn't have tests, write them first.

```
claude "Before we refactor src/services/billing-service.ts,
  write comprehensive tests for its current behavior. Cover:
  - All public methods
  - Edge cases (empty inputs, null values, boundary conditions)
  - Error cases
  - The current behavior, not the desired behavior

  These tests will serve as a safety net during refactoring."
```

This is one of AI's highest-value contributions to refactoring: writing characterization tests that capture current behavior, even when that behavior is messy. These tests ensure the refactor doesn't accidentally change what the code does.

Step 2: Refactor in Small, Testable Steps

Don't ask AI to refactor everything at once. Break it into steps:

```
## Step 1: Extract the validation logic
claude "Extract the input validation from processOrder into a
  separate validateOrderInput function. Keep everything else the same."

## Run tests. All pass? Good. Commit.

## Step 2: Extract the pricing logic
claude "Extract the pricing calculation from processOrder into a
  separate calculateOrderPricing function."

## Run tests. All pass? Good. Commit.

## Step 3: Extract the persistence logic
claude "Extract the database operations from processOrder into a
  separate persistOrder function."

## Run tests. All pass? Commit. Now processOrder is a clean
## composition of three well-named functions.
```

Each step is small, testable, and independently committable. If something breaks, you know which step caused it.

Step 3: Verify Behavior Preservation

After AI finishes a refactor, verify:

1. All existing tests pass
2. The external interface hasn't changed (same function signatures, same API contracts)
3. The behavior is identical (run the application and test manually)
4. No new dependencies were introduced unnecessarily

```
## After refactoring, run your full test suite
npm test

## Check for interface changes
```

```
git diff --stat  # See which files changed
git diff -- src/services/billing-service.ts  # Review specific file
```

Common Refactoring Patterns with AI

Pattern 1: Replace Conditionals with Polymorphism

AI handles this classic refactor well:

```
This function has a massive switch statement that handles
different notification types. Refactor it to use the Strategy
pattern – create a NotificationStrategy interface and separate
classes for each notification type (EmailNotification,
SMSNotification, PushNotification, SlackNotification).
```

Before:

```
function sendNotification(type: string, recipient: string, message: string) {
  switch (type) {
    case 'email':
      // 30 lines of email-specific code
      break;
    case 'sms':
      // 25 lines of SMS-specific code
      break;
    case 'push':
      // 20 lines of push notification code
      break;
    case 'slack':
      // 35 lines of Slack-specific code
      break;
    default:
      throw new Error(`Unknown notification type: ${type}`);
  }
}
```

After:

```
interface NotificationStrategy {
  send(recipient: string, message: string): Promise<void>;
}

class EmailNotification implements NotificationStrategy {
  async send(recipient: string, message: string): Promise<void> {
    // Email-specific logic, clean and isolated
```

```
  }
}

// ... similar classes for SMS, Push, Slack

const strategies: Record<string, NotificationStrategy> = {
  email: new EmailNotification(),
  sms: new SMSNotification(),
  push: new PushNotification(),
  slack: new SlackNotification(),
};

async function sendNotification(
  type: string,
  recipient: string,
  message: string
): Promise<void> {
  const strategy = strategies[type];
  if (!strategy) throw new Error(`Unknown notification type: ${type}`);
  await strategy.send(recipient, message);
}
```

Pattern 2: Dependency Injection

Replacing hardcoded dependencies with injected ones — essential for testability.

```
Refactor UserService to use dependency injection. Currently it
creates its own database connection and email client internally.
Change it to accept these as constructor parameters. Update all
call sites. Update tests to inject mocks instead of mocking modules.
```

Pattern 3: API Response Standardization

When your API has inconsistent response formats:

```
claude "Our API responses are inconsistent. Some endpoints return
  { data: ... }, some return the data directly, some return
  { result: ..., success: true }.

  Standardize ALL API responses to use this envelope:
  { data: T, error: null, meta: { timestamp, requestId } } for success
  { data: null, error: { code, message }, meta: { timestamp, requestId } } for e
rrors
```

```
  Update all route handlers in src/routes/. Update the response
  type definitions. Update tests to expect the new format."
```

Pattern 4: Module Boundary Cleanup

When modules have become entangled:

```
claude "Analyze the import graph of src/services/. Identify
  circular dependencies and suggest how to break them. For each
  circular dependency, explain which module should own the shared
  code and how to restructure the imports."
```

Tackling Legacy Code

Legacy code — no tests, outdated patterns, written by people who left the company — is where AI refactoring provides the most value and carries the most risk.

The Legacy Code Strategy

1. **Understand first.** Ask AI to explain the code before changing it.

```
claude "Explain what src/legacy/order-processor.js does.
  This is a 500-line file with no tests and no documentation.
  Walk me through the main logic flow, the side effects,
  and any potential issues you see."
```

2. **Add types.** If it's JavaScript, add TypeScript types. If it's Python, add type hints. This surfaces bugs immediately and makes future changes safer.

```
claude "Convert src/legacy/order-processor.js to TypeScript.
  Add proper types for all functions, parameters, and return values.
  Don't change any logic - just add types. Flag any places where
  the types reveal potential bugs."
```

3. **Add characterization tests.** Tests that capture current behavior, bugs and all.

4. **Refactor incrementally.** Small, safe steps with tests running after each one.

What AI Gets Wrong with Legacy Code

- **Assuming modern patterns:** AI may refactor to async/await in a codebase that can't support it due to other dependencies
- **Breaking implicit contracts:** Legacy code often has undocumented behaviors that other code depends on
- **Optimistic simplification:** AI may remove "dead" code that's actually used via reflection, dynamic requires, or configuration-driven loading
- **Ignoring deployment constraints:** A refactor that requires a database migration in a zero-downtime environment needs careful planning AI doesn't do automatically

Always check AI's refactoring suggestions against your production deployment constraints.

Measuring Technical Debt Reduction

Justifying refactoring time to stakeholders is hard without metrics. AI can help you quantify technical debt:

```
claude "Analyze this codebase and produce a technical debt report:
1. Functions exceeding 50 lines (count and list)
2. Files exceeding 500 lines (count and list)
3. Circular dependencies (count and list)
4. Functions with cyclomatic complexity > 10
5. Modules with no test coverage
6. Deprecated API usage
7. TODO/FIXME/HACK comments (count and categorize)

Format as a markdown table with severity ratings."
```

Run this analysis monthly. Track the numbers. Show stakeholders the trend line. "Technical debt items reduced from 147 to 89 over three months" tells a compelling story.

The Refactoring Workflow in Practice

Here's a practical example. You have a 400-line Express route handler for order creation. It validates input, checks inventory, calculates pricing, processes payment, sends confirmation emails, and updates analytics. All in one function.

Step 1: Characterize

```
claude "Read src/routes/orders.ts, specifically the POST / handler.
  Write characterization tests that capture its current behavior
  for: valid orders, invalid input, out-of-stock items,
  payment failures, and partial failures."
```

Step 2: Extract in order

```
claude "Extract the input validation logic from the POST / handler
  into a validateCreateOrderRequest function in src/validators/."
```

Run tests. Commit.

```
claude "Extract the inventory checking logic into an
  InventoryService.checkAvailability method."
```

Run tests. Commit.

Continue until the route handler is a clean 20-line orchestration function.

Step 3: Verify

```
claude "Compare the behavior of the refactored POST /orders
  endpoint with the original. Are there any behavioral differences?
  Run the characterization tests and report results."
```

This systematic approach turns a terrifying 400-line monster into clean, testable, maintainable modules — AI does the heavy lifting, tests make sure nothing breaks.

Refactoring makes existing code better. Testing keeps it that way. Chapter 8 covers using AI for test generation — unit tests to integration tests, TDD workflows to coverage strategies. If you've been skipping tests because they take too long to write, AI just removed your last excuse.

Chapter 8: Testing with AI

Test Generation, TDD with AI, Coverage Strategies

Testing is the part of software development everyone agrees is important and nobody wants to do. It's tedious, repetitive, and feels like it slows you down — even though you know it speeds you up over time.

AI removes the tedium. Generating tests with AI is fast, thorough, and surprisingly enjoyable. AI-generated test suites routinely cover edge cases humans skip because writing another test case feels like diminishing returns. With AI, there are no diminishing returns — the 50th test case costs the same effort as the first.

This chapter covers generating tests across multiple frameworks, using AI for TDD, building coverage strategies, and the critical question of when AI-generated tests are actually useful versus when they're testing theater.

The Test Generation Workflow

Unit Tests from Existing Code

The most common pattern: you have code, you need tests.

```
claude "Write unit tests for src/services/pricing-service.ts.
  Use Vitest. Cover:
  - All public methods
  - Happy path for each method
  - Edge cases: empty inputs, zero values, negative values
  - Error cases: invalid inputs, missing dependencies
  - Boundary conditions: max values, min values, floating point

  Follow the testing patterns in our existing test files
  (see src/services/__tests__/ for examples)."
```

AI will generate comprehensive test suites. Here's an example of what you get:

```
import { describe, it, expect, beforeEach, vi } from 'vitest';
import { PricingService } from '../pricing-service';
import { TaxService } from '../tax-service';
```

```
import { DiscountEngine } from '../discount-engine';

describe('PricingService', () => {
  let pricingService: PricingService;
  let mockTaxService: TaxService;
  let mockDiscountEngine: DiscountEngine;

  beforeEach(() => {
    mockTaxService = {
      calculateTax: vi.fn().mockResolvedValue(new Decimal('8.00')),
      getTaxRate: vi.fn().mockReturnValue(new Decimal('0.08')),
    } as unknown as TaxService;

    mockDiscountEngine = {
      applyDiscounts: vi.fn().mockResolvedValue(new Decimal('0')),
    } as unknown as DiscountEngine;

    pricingService = new PricingService(mockTaxService, mockDiscountEngine);
  });

  describe('calculateTotal', () => {
    it('should calculate total for a single item', async () => {
      const items = [{ productId: 'p1', quantity: 2, unitPrice: new Decimal('25.
00') }];
      const result = await pricingService.calculateTotal(items);

      expect(result.subtotal).toEqual(new Decimal('50.00'));
      expect(result.tax).toEqual(new Decimal('8.00'));
      expect(result.total).toEqual(new Decimal('58.00'));
    });

    it('should handle empty item list', async () => {
      const result = await pricingService.calculateTotal([]);

      expect(result.subtotal).toEqual(new Decimal('0'));
      expect(result.tax).toEqual(new Decimal('0'));
      expect(result.total).toEqual(new Decimal('0'));
    });

    it('should throw on negative quantity', async () => {
      const items = [{ productId: 'p1', quantity: -1, unitPrice: new Decimal('25.
00') }];

      await expect(pricingService.calculateTotal(items))
        .rejects.toThrow('Quantity must be positive');
```

```
  });

  it('should throw on zero unit price', async () => {
    const items = [{ productId: 'p1', quantity: 1, unitPrice: new Decimal('0')
}];

    await expect(pricingService.calculateTotal(items))
      .rejects.toThrow('Unit price must be positive');
  });

  it('should apply discounts before tax', async () => {
    mockDiscountEngine.applyDiscounts = vi.fn()
      .mockResolvedValue(new Decimal('10.00'));

    const items = [{ productId: 'p1', quantity: 1, unitPrice: new Decimal('100.
00') }];
    const result = await pricingService.calculateTotal(items);

    expect(result.discount).toEqual(new Decimal('10.00'));
    expect(mockTaxService.calculateTax).toHaveBeenCalledWith(
      new Decimal('90.00') // Tax on discounted amount
    );
  });
 });
});
```

Tests from Requirements

Even better: generate tests before the code exists.

```
I need to implement a password strength validator.
Write the tests first (TDD style) using Vitest.

Requirements:
- Minimum 8 characters
- At least one uppercase letter
- At least one lowercase letter
- At least one digit
- At least one special character (!@#$%^&*)
- No more than 3 consecutive identical characters
- Not in a list of common passwords
- Returns a strength score: weak, fair, strong, very_strong
- Returns specific feedback messages for each failed rule
```

```
Write comprehensive tests, then I'll implement the function
to make them pass.
```

This is AI-assisted TDD. You get a complete test suite that serves as both spec and verification. Then you (or AI) write the implementation to make the tests pass.

Integration Tests

AI generates integration tests that test real interactions:

```python
## Prompt: "Write integration tests for the order creation API endpoint"

import pytest
from httpx import AsyncClient
from app.main import app
from app.database import get_test_db

@pytest.fixture
async def client():
    async with AsyncClient(app=app, base_url="http://test") as client:
        yield client

@pytest.fixture
async def authenticated_client(client):
    response = await client.post("/auth/login", json={
        "email": "test@example.com",
        "password": "testpass123"
    })
    token = response.json()["data"]["token"]
    client.headers["Authorization"] = f"Bearer {token}"
    return client

@pytest.fixture
async def sample_product(db):
    product = await db.products.create({
        "name": "Test Widget",
        "price": "29.99",
        "stock": 100
    })
    yield product
    await db.products.delete(product.id)

class TestCreateOrder:
```

```
async def test_creates_order_successfully(
    self, authenticated_client, sample_product
):
    response = await authenticated_client.post("/api/orders", json={
        "items": [{"product_id": sample_product.id, "quantity": 2}]
    })

    assert response.status_code == 201
    data = response.json()["data"]
    assert data["status"] == "pending"
    assert len(data["items"]) == 1
    assert data["total"] == "59.98"

async def test_rejects_unauthenticated_request(self, client):
    response = await client.post("/api/orders", json={
        "items": [{"product_id": "p1", "quantity": 1}]
    })
    assert response.status_code == 401

async def test_rejects_empty_order(self, authenticated_client):
    response = await authenticated_client.post("/api/orders", json={
        "items": []
    })
    assert response.status_code == 400
    assert "at least one item" in response.json()["error"]["message"].lower()

async def test_rejects_out_of_stock_item(
    self, authenticated_client, sample_product, db
):
    await db.products.update(sample_product.id, {"stock": 0})

    response = await authenticated_client.post("/api/orders", json={
        "items": [{"product_id": sample_product.id, "quantity": 1}]
    })
    assert response.status_code == 409
    assert "out of stock" in response.json()["error"]["message"].lower()

async def test_handles_concurrent_orders(
    self, authenticated_client, sample_product, db
):
    """Two orders for the last item — one should succeed, one should fail.""
"

    await db.products.update(sample_product.id, {"stock": 1})

    import asyncio
```

```
    results = await asyncio.gather(
        authenticated_client.post("/api/orders", json={
            "items": [{"product_id": sample_product.id, "quantity": 1}]
        }),
        authenticated_client.post("/api/orders", json={
            "items": [{"product_id": sample_product.id, "quantity": 1}]
        }),
        return_exceptions=True
    )

    statuses = [r.status_code for r in results]
    assert 201 in statuses  # One succeeds
    assert 409 in statuses  # One fails (out of stock)
```

Notice that AI generated a concurrency test — something most developers skip because it's hard to write manually.

TDD with AI

Test-driven development works naturally with AI. The workflow:

1. **You write the test (or describe what to test, and AI writes it)**
2. **AI implements the code to make the test pass**
3. **You review both the test and the implementation**
4. **Refactor as needed**

```
## Step 1: Generate the test
claude "Write a failing test for a function `parseCSVRow` that:
  - Takes a CSV row string and a header array
  - Returns an object mapping headers to values
  - Handles quoted values with commas inside
  - Handles escaped quotes
  - Throws on mismatched column count
  Use Vitest."

## Step 2: Implement to pass the test
claude "Now implement parseCSVRow to make all those tests pass."

## Step 3: Review and refactor
claude "Review the implementation. Is the CSV parsing correct
  for all edge cases? Are there any cases the tests miss?"
```

Coverage Strategies

Identifying Coverage Gaps

```
## After running tests with coverage
claude "Here's our test coverage report. We're at 73% overall.
  The following files have less than 50% coverage:
  - src/services/notification-service.ts (23%)
  - src/utils/date-helpers.ts (41%)
  - src/middleware/rate-limiter.ts (12%)

  Prioritize which to test first based on risk (how critical is
  the code) and generate tests for the highest-priority file."
```

The 80/20 Rule of Test Coverage

Don't blindly chase 100% coverage. AI can generate tests for every line of code, but not all tests are equally valuable.

High-value tests: - Business logic with complex rules - Functions that handle money, authentication, or authorization - Code paths that handle errors and edge cases - Integration points with external services

Low-value tests: - Simple getters/setters - Framework boilerplate - Configuration files - UI layout code (test behavior, not pixels)

```
claude "Analyze our codebase and identify the highest-risk
  untested code paths. Rank by: business impact if this code
  breaks, complexity of the logic, and how often this code
  is modified. Generate tests for the top 5."
```

When AI-Generated Tests Are Bad

AI-generated tests can give you a false sense of security. Watch for these anti-patterns:

Testing the Implementation, Not the Behavior

```
// BAD: This test is coupled to the implementation
it('should call database.query with the correct SQL', () => {
  getUser('123');
  expect(db.query).toHaveBeenCalledWith(
    'SELECT * FROM users WHERE id = $1', ['123']
```

```
  );
});

// GOOD: This test checks behavior, not implementation
it('should return the user with the given ID', async () => {
  await db.users.create({ id: '123', name: 'Alice' });
  const user = await getUser('123');
  expect(user.name).toBe('Alice');
});
```

AI frequently generates the first kind because it's easier — just verify the mock was called correctly. Push back and ask for behavior tests.

Testing That the Code Does What the Code Does

```
// BAD: This just restates the implementation as a test
it('should return price * quantity', () => {
  expect(calculateLineTotal(10, 3)).toBe(30);
});
// This test adds no value — it will pass even if the business
// rule should be price * quantity * tax_rate
```

Good tests encode business requirements, not code logic. Tell AI about the business rules explicitly.

Over-Mocking

```
// BAD: Everything is mocked, so what are you actually testing?
it('should process order', async () => {
  mockInventory.check.mockResolvedValue(true);
  mockPricing.calculate.mockResolvedValue(100);
  mockPayment.charge.mockResolvedValue({ success: true });
  mockNotification.send.mockResolvedValue(undefined);

  const result = await processOrder(mockOrder);

  expect(result.success).toBe(true);
  // This test passes even if processOrder is completely wrong
  // because every dependency returns a canned success
});
```

If every dependency is mocked, you're testing that function calls happen in order — not that they produce correct results. Use real implementations where practical.

Multi-Language Test Generation

Go Table-Driven Tests

```go
// Prompt: "Generate table-driven tests for this Go function"

func TestParseVersion(t *testing.T) {
    tests := []struct {
        name    string
        input   string
        want    Version
        wantErr bool
    }{
        {
            name:  "standard version",
            input: "1.2.3",
            want:  Version{Major: 1, Minor: 2, Patch: 3},
        },
        {
            name:  "version with v prefix",
            input: "v1.2.3",
            want:  Version{Major: 1, Minor: 2, Patch: 3},
        },
        {
            name:  "major only",
            input: "5",
            want:  Version{Major: 5, Minor: 0, Patch: 0},
        },
        {
            name:    "empty string",
            input:   "",
            wantErr: true,
        },
        {
            name:    "non-numeric",
            input:   "abc",
            wantErr: true,
        },
        {
            name:    "negative version",
            input:   "-1.0.0",
            wantErr: true,
        },
    }
```

```go
    for _, tt := range tests {
        t.Run(tt.name, func(t *testing.T) {
            got, err := ParseVersion(tt.input)
            if tt.wantErr {
                if err == nil {
                    t.Errorf("ParseVersion(%q) expected error, got %v", tt.input,
 got)
                }
                return
            }
            if err != nil {
                t.Errorf("ParseVersion(%q) unexpected error: %v", tt.input, err)
                return
            }
            if got != tt.want {
                t.Errorf("ParseVersion(%q) = %v, want %v", tt.input, got, tt.wan
t)
            }
        })
    }
}
```

Python Parametrized Tests

```python
## Prompt: "Generate pytest parametrized tests for the email validator"

import pytest
from validators import validate_email

@pytest.mark.parametrize("email,expected", [
    ("user@example.com", True),
    ("user.name@example.com", True),
    ("user+tag@example.com", True),
    ("user@subdomain.example.com", True),
    ("", False),
    ("not-an-email", False),
    ("@no-local-part.com", False),
    ("no-domain@", False),
    ("spaces in@email.com", False),
    ("user@.com", False),
    ("user@com", False),  # Debatable — check your business rules
    ("a" * 255 + "@example.com", False),  # Too long
])
```

```python
def test_validate_email(email: str, expected: bool):
    assert validate_email(email) == expected, f"Failed for: {email!r}"
```

The Testing Flywheel

AI creates a positive feedback loop for testing:

1. AI generates tests quickly → you write more tests
2. More tests → higher confidence in refactoring
3. Higher confidence → more refactoring with AI
4. More refactoring → cleaner code → easier to test
5. Easier to test → AI generates even better tests

The flywheel is real. Teams that adopt AI testing consistently report that test coverage increases over time without additional effort — because the marginal cost of adding a test approaches zero.

Chapter 9: Documentation That Writes Itself

READMEs, API Docs, Inline Comments, Architecture Docs

Documentation is the most thankless task in software development. Nobody wants to write it, everybody wants it to exist, and it goes stale the moment you commit it.

AI changes the economics. What used to take 30 minutes now takes 3 minutes. What used to require a dedicated documentation sprint now happens alongside development. The quality isn't perfect — AI documentation needs editing — but a good first draft you edit is infinitely better than the blank page you never fill.

This chapter covers generating every type of documentation you need, from inline comments to architecture decision records, across every major AI tool.

Inline Comments and Docstrings

The lowest-friction documentation: comments that live in your code.

Generating Docstrings

```
claude "Add comprehensive docstrings to all public functions
  and classes in src/services/. Follow the existing docstring
  style in the project. Include parameter descriptions, return
  values, exceptions thrown, and usage examples where helpful."
```

AI generates thorough, consistent docstrings:

```
class InventoryService:
    """Manages product inventory levels and availability checks.

    This service handles all inventory-related operations including
    stock checks, reservations, and adjustments. It coordinates with
    the warehouse system for physical inventory counts.

    All methods are transactional — partial updates are rolled back
    on failure.
```

```python
    Example:
        service = InventoryService(db_session, warehouse_client)
        available = await service.check_availability("SKU-001", quantity=5)
        if available:
            reservation = await service.reserve("SKU-001", quantity=5,
                                                ttl_minutes=15)
    """

    async def check_availability(
        self,
        sku: str,
        quantity: int,
        warehouse_id: str | None = None
    ) -> bool:
        """Check if a product is available in the requested quantity.

        Args:
            sku: Product SKU identifier.
            quantity: Number of units needed. Must be positive.
            warehouse_id: Optional warehouse to check. If None, checks
                all warehouses.

        Returns:
            True if the requested quantity is available, False otherwise.

        Raises:
            ProductNotFoundError: If the SKU doesn't exist.
            ValueError: If quantity is not positive.
        """
```

TypeScript JSDoc

```typescript
/**
 * Merges two user profile objects with configurable precedence.
 *
 * Non-null values from the secondary profile override the primary.
 * Arrays are merged and deduplicated. Nested objects are deep-merged.
 *
 * @param primary - The base profile (provides defaults and ID)
 * @param secondary - The overriding profile (non-null values take precedence)
 * @returns A new merged UserProfile (inputs are not mutated)
 *
 * @example
 * ```typescript
```

```
 * const merged = mergeProfiles(existingProfile, updatedFields);
 * // merged.id === existingProfile.id (always from primary)
 * // merged.name === updatedFields.name (if not null)
 * ```
 *
 * @throws {InvalidProfileError} If either profile has an invalid structure
 */
export function mergeProfiles(
  primary: UserProfile,
  secondary: Partial<UserProfile>
): UserProfile {
```

When NOT to Comment

AI will happily add comments to every line of code. Push back on noise:

```
## BAD — AI-generated noise comments
x = x + 1  # Increment x by 1
users = db.get_users()  # Get users from database
if user is None:  # Check if user is None
    return None  # Return None

## GOOD — comments explain WHY, not WHAT
## Offset by 1 because the API uses 1-based pagination
x = x + 1

## Cache miss — fall through to database query
users = db.get_users()
```

Tell AI explicitly: “Only add comments that explain WHY, not WHAT. Do not restate what the code already says.”

README Generation

Project README

```
claude "Generate a comprehensive README.md for this project. Include:
- Project name and one-line description
- Status badges placeholder
- Prerequisites and installation steps
- Quick start guide with a working example
- Configuration reference
- Development setup (how to run locally)
- Testing instructions
```

```
- Deployment overview
- Contributing guidelines summary
- License

Base everything on the actual project structure, package.json,
and existing code. Do not make up features that don't exist."
```

That last instruction matters. AI READMEs tend to describe the project you wish you had, not the one you have. Explicitly constraining to actual features prevents that.

Module-Level READMEs

For monorepos or complex projects, generate READMEs for individual modules:

```
claude "Generate a README for the src/services/ directory.
  Explain what services exist, their responsibilities,
  how they interact, and how to add a new service following
  our existing patterns."
```

API Documentation

OpenAPI/Swagger from Code

AI generates OpenAPI specifications from your route handlers:

```
claude "Generate an OpenAPI 3.1 specification for our Express API.
  Read all route files in src/routes/ and extract:
  - Endpoints (method + path)
  - Request parameters (path, query, body)
  - Request body schemas (from Zod validators)
  - Response schemas (from actual response shapes)
  - Authentication requirements
  - Error responses

  Output as openapi.yaml in the project root."
```

The output:

```yaml
openapi: 3.1.0
info:
  title: Inventory Management API
  version: 1.0.0
  description: API for managing product inventory and orders
```

```yaml
paths:
  /api/products:
    get:
      summary: List all products
      parameters:
        - name: page
          in: query
          schema:
            type: integer
            default: 1
        - name: limit
          in: query
          schema:
            type: integer
            default: 20
            maximum: 100
      responses:
        '200':
          description: Paginated list of products
          content:
            application/json:
              schema:
                type: object
                properties:
                  data:
                    type: array
                    items:
                      $ref: '#/components/schemas/Product'
                  meta:
                    $ref: '#/components/schemas/PaginationMeta'
    post:
      summary: Create a new product
      security:
        - bearerAuth: []
      requestBody:
        required: true
        content:
          application/json:
            schema:
              $ref: '#/components/schemas/CreateProductRequest'
      responses:
        '201':
          description: Product created successfully
        '400':
          description: Validation error
```

```yaml
        '401':
          description: Authentication required
```

API Reference Pages

For human-readable API documentation:

```
claude "Generate API reference documentation for each endpoint
  in markdown format. For each endpoint include:
  - URL and method
  - Description of what it does
  - Authentication requirements
  - Request parameters with types and descriptions
  - Example request (curl)
  - Example response (JSON)
  - Error responses with codes and messages
  - Rate limiting information

  Write one markdown file per resource (users, products, orders)."
```

Architecture Documentation

Architecture Decision Records (ADRs)

AI turns conversations about design decisions into ADRs:

```
Write an Architecture Decision Record for the following decision:

Context: We need to add real-time notifications to our application.
Users need to see inventory alerts, order status changes, and
system announcements without refreshing.

Options considered:
1. WebSockets (Socket.IO)
2. Server-Sent Events (SSE)
3. Polling (every 5 seconds)

Decision: We chose SSE because:
- Our notifications are server-to-client only (no bidirectional need)
- SSE has automatic reconnection built into the browser API
- SSE works through HTTP proxies and load balancers without special config
- Lower complexity than WebSockets for our use case
```

Use the standard ADR format (Title, Status, Context, Decision, Consequences).

Output:

```markdown
## ADR-007: Use Server-Sent Events for Real-Time Notifications

### Status
Accepted (2026-03-15)

### Context
The application needs real-time notification delivery for inventory
alerts, order status changes, and system announcements. Users currently
must refresh the page to see updates, causing missed alerts and a
poor user experience.

Three approaches were evaluated:
1. **WebSockets (Socket.IO)** — Full bidirectional communication
2. **Server-Sent Events (SSE)** — Server-to-client streaming over HTTP
3. **HTTP Polling** — Client polls for updates every 5 seconds

### Decision
We will use **Server-Sent Events (SSE)** for real-time notification
delivery.

#### Rationale
- Notifications are exclusively server-to-client; bidirectional
  communication is not required
- SSE provides automatic reconnection via the native EventSource API
- SSE operates over standard HTTP, requiring no special proxy or
  load balancer configuration
- Implementation complexity is significantly lower than WebSockets
- SSE is supported by all modern browsers

#### Trade-offs Accepted
- If we later need client-to-server real-time communication
  (e.g., collaborative editing), SSE will be insufficient
- SSE has a limit of ~6 concurrent connections per domain in
  HTTP/1.1 (not an issue with HTTP/2)

### Consequences
- Add an SSE endpoint at `/api/notifications/stream`
- Implement a server-side event bus for publishing notifications
- Client-side: use native EventSource API (no library needed)
```

```
- Add connection management and graceful degradation to polling
  for environments where SSE is blocked
```

System Architecture Overviews

```
claude "Read the entire codebase and generate an architecture
  overview document. Include:
  - High-level system diagram (describe in text/Mermaid format)
  - Component descriptions
  - Data flow between components
  - External dependencies and integrations
  - Key design patterns used
  - Deployment architecture

  Base this on actual code, not assumptions. If something is
  unclear from the code, say so."
```

AI generates Mermaid diagrams that render in GitHub and documentation sites:

```
graph TB
    Client[React Frontend] -->|REST API| Gateway[API Gateway]
    Gateway -->|Auth| Auth[Auth Service]
    Gateway -->|Routes| API[Express API]
    API --> Services[Service Layer]
    Services --> Repos[Repository Layer]
    Repos --> DB[(PostgreSQL)]
    Services -->|Events| Queue[Redis Pub/Sub]
    Queue --> Workers[Background Workers]
    Workers --> Email[Email Service]
    Workers --> Analytics[Analytics Pipeline]
```

Changelog Generation

AI generates changelogs from your git history:

```
claude "Generate a CHANGELOG entry for the v2.4.0 release.
  Read the git log from the v2.3.0 tag to HEAD.
  Group changes into:
  - Features
  - Bug Fixes
  - Performance Improvements
  - Breaking Changes
  - Dependencies

  Use the Keep a Changelog format. Be concise but descriptive."
```

```
### [2.4.0] - 2026-03-28

#### Features
- Add bulk CSV import for inventory items with validation preview
- Add SSE-based real-time notifications for inventory alerts
- Add product image upload with automatic thumbnail generation

#### Bug Fixes
- Fix race condition in concurrent order processing (#247)
- Fix incorrect tax calculation for multi-state orders (#251)
- Fix memory leak in WebSocket connection cleanup (#253)

#### Performance
- Reduce product search query time by 60% with GIN index
- Add Redis caching for frequently accessed product data

#### Breaking Changes
- API response envelope changed: `result` renamed to `data`
- Minimum Node.js version raised to 20 LTS

#### Dependencies
- Upgrade Prisma from 5.8 to 5.12
- Upgrade React from 18.2 to 18.3
```

Documentation Maintenance

The hardest part of documentation is keeping it current. AI helps here too.

Drift Detection

```
claude "Compare our API documentation (docs/api/) with the actual
  API routes (src/routes/). Identify any discrepancies:
  - Endpoints documented but not implemented
  - Endpoints implemented but not documented
  - Parameter mismatches
  - Response format differences

  Report all discrepancies."
```

Automated Doc Updates

After modifying code, update the documentation:

```
claude "I just added a `bulkUpdate` method to ProductService.
  Update:
  1. The ProductService JSDoc
  2. The API documentation in docs/api/products.md
  3. The OpenAPI spec
  4. The README's API overview section

  Make sure all four are consistent."
```

Documentation Quality Principles

Even with AI, documentation quality requires human judgment. Apply these principles:

Write for the reader, not the writer. AI sometimes describes internals the reader doesn't care about. Ask: "Would a developer using this API need to know this?"

Examples beat explanations. A working code example is worth a thousand words of prose. Always include examples.

Keep it DRY. AI will happily duplicate information across files. Centralize and reference.

Date it. Documentation without a date is documentation you can't trust. Add "Last updated:" headers.

Review AI-generated docs. AI occasionally invents features, mischaracterizes behavior, or describes aspirational functionality rather than actual. Read every word.

The Documentation Workflow

Integrate documentation into your development flow:

1. **New feature:** Generate API docs and update README as part of the PR
2. **Bug fix:** Update relevant docs if the fix changes expected behavior
3. **Refactoring:** Regenerate architecture docs after significant restructuring
4. **Release:** Generate changelog from git history

With AI, documentation is no longer a separate activity. It's a natural part of writing code. The excuse "we don't have time for docs" stops holding when docs take 3 minutes instead of 30.

From the What to the Why

Documentation explains the what and the how. Architecture explains the why. Chapter 10 covers using AI for architecture and system design — the high-level decisions that determine whether your system succeeds or fails at scale. This is where AI shifts from a code tool to a thinking partner.

Part III: Architecture and Domains

"There are only two hard things in Computer Science: cache invalidation and naming things."

— Phil Karlton

Chapter 10: Architecture and System Design

Using AI for Design Decisions, Trade-Off Analysis, RFC Drafting

Architecture is where software engineering is most human. It requires weighing incomplete information, predicting future needs, balancing competing concerns, and making decisions you'll live with for years. AI can't replace architectural judgment. But it can be the best thinking partner you've ever had.

The key insight: AI is not good at making architectural decisions. It's excellent at helping you explore the decision space — surfacing options you hadn't considered, articulating trade-offs you felt but couldn't name, and stress-testing your assumptions before you commit.

AI as Architecture Sounding Board

The most valuable use of AI in architecture is the conversation. Not "design my system" but "help me think about my system."

Exploring Options

```
I'm designing a notification system for a SaaS product.
Users need to receive notifications via email, in-app,
SMS, and Slack. Some notifications are urgent (security alerts),
some are routine (weekly digests). The system handles about
100K notifications per day now and might grow to 1M.

What architectural approaches should I consider? For each one,
give me the pros, cons, and when it's the right choice.
```

AI will typically outline several approaches:

1. **Direct send from application code** — Simple but tightly coupled
2. **Event-driven with a message queue** — Decoupled, scalable, more complex

3. **Dedicated notification microservice** — Clean separation but operational overhead
4. **Third-party notification service (Novu, Knock)** — Fast to implement, dependency risk

For each, AI provides detailed trade-off analysis that would take you an hour to research and articulate yourself. It draws from patterns it's seen across thousands of systems.

Stress-Testing a Design

Once you have a design in mind, use AI to attack it.

```
Here's my proposed architecture for the notification system:
[describe architecture]

Play devil's advocate. What are the failure modes? What happens
when the message queue goes down? What happens under 10x traffic
spike? Where are the single points of failure? What will I regret
in 12 months?
```

This is enormously valuable. AI will surface failure modes you hadn't considered, because it has seen them in other systems. Common findings:

- **Queue poison messages:** What happens when a notification fails to send? Does it retry forever?
- **Ordering guarantees:** Do users receive notifications in the order they were triggered?
- **Idempotency:** If a message is delivered twice from the queue, does the user get two emails?
- **Backpressure:** When the email provider throttles you, does the queue back up and consume memory?

Comparing Approaches

```
Compare these two approaches for our user permissions system:

Approach A: Role-Based Access Control (RBAC)
- Users have roles (admin, editor, viewer)
- Roles have permissions
- Permissions checked at route level
```

```
Approach B: Attribute-Based Access Control (ABAC)
- Policies define access based on user attributes,
  resource attributes, and context
- More flexible but more complex

Our application: multi-tenant SaaS with ~50 distinct permission
scenarios. Tenants want to customize their own roles. We have
3 developers and ship weekly.

Which approach is better for us and why?
```

AI will analyze this against your specific constraints — team size, complexity budget, the customization requirement — and likely recommend starting with RBAC augmented with a few attribute-based rules rather than going full ABAC. The answer is context-dependent, and good AI analysis respects that context.

A Concrete Example: RFC from Scratch

Let me walk you through what this actually looks like. You've been asked to design an authentication system for a new microservices platform. Instead of staring at a blank document, you open a conversation with AI.

Your prompt:

```
We're migrating from a monolith to microservices and need to design
the authentication architecture. Current state:
- Monolith uses session cookies stored in Redis
- 50K daily active users
- 12 microservices planned (3 built so far)
- Services communicate over gRPC internally
- External API uses REST

We need to support:
- User authentication (email/password, Google OAuth, SAML for enterprise)
- Service-to-service authentication
- API key authentication for third-party integrations
- Fine-grained permissions per service

Write an RFC for this. Our RFC template uses: Summary, Motivation,
Detailed Design, Security Considerations, Alternatives Considered,
Migration Plan, and Open Questions.
```

AI produces a 3-page RFC draft. The interesting part isn't that it generates text — it's what it includes that you might not have thought of:

- **Token format decision:** It lays out JWT vs. opaque tokens with specific trade-offs for your gRPC service mesh (JWTs avoid a network hop to the auth service on every request, but you can't revoke them instantly).
- **Service identity:** It proposes mTLS for service-to-service auth, which is the right call for gRPC but something teams often defer until "later."
- **Migration strategy:** It suggests running session cookies and JWTs in parallel during migration, with middleware that accepts both.

Here's the part you need to edit. AI's draft includes an "Alternatives Considered" section that dismisses API gateway authentication too quickly. In your case, you already have Kong deployed, and offloading auth to the gateway for external requests is the right call. AI doesn't know your infrastructure. You rewrite that section with your actual context.

The RFC goes from blank page to reviewable draft in 15 minutes instead of half a day. Your team reviews it, argues about the JWT revocation strategy (as they should), and you update the RFC based on the discussion. AI got you to the starting line of a productive conversation faster. That's the value.

What AI gets wrong in RFCs: Watch for generic recommendations that ignore your constraints. AI will suggest "use an API gateway" or "use a service mesh" without knowing whether you already have one, whether your team can operate it, or whether your scale justifies it. Always map AI's recommendations against your actual infrastructure and team capabilities.

RFC and Design Document Generation

Writing RFCs from Conversations

After an architecture discussion, you often need to formalize the decision. AI excels at turning a conversation into a structured document:

```
Based on our discussion, write an RFC for the notification system.

Use this format:
- Title
- Authors
```

```
- Status (Draft)
- Summary (2-3 sentences)
- Motivation (why we need this)
- Detailed Design (the actual architecture)
- Alternatives Considered (and why we rejected them)
- Rollout Plan (how we'll deploy this incrementally)
- Open Questions (things we still need to decide)

Include a Mermaid diagram of the architecture.
```

AI generates a complete RFC draft:

````markdown
## RFC: Notification Service Architecture

**Authors:** [Your Name]
**Status:** Draft
**Date:** 2026-03-28

### Summary
This RFC proposes an event-driven notification service that
decouples notification logic from application code, supports
multiple delivery channels, and scales to 1M+ notifications/day.

### Motivation
The current notification implementation is scattered across
the application codebase...

### Detailed Design

#### Architecture Overview

```mermaid
graph LR
    App[Application] -->|Events| Bus[Event Bus]
    Bus --> Router[Notification Router]
    Router --> Email[Email Worker]
    Router --> InApp[In-App Worker]
    Router --> SMS[SMS Worker]
    Router --> Slack[Slack Worker]
    Email --> SES[AWS SES]
    SMS --> Twilio[Twilio]
    Slack --> SlackAPI[Slack API]
    InApp --> DB[(Notification DB)]
    DB --> SSE[SSE Endpoint]
````
````

Event Schema

```
interface NotificationEvent {
  id: string;                    // Idempotency key
  type: 'security_alert' | 'order_update' | 'weekly_digest';
  recipientId: string;
  channels: ('email' | 'in_app' | 'sms' | 'slack')[];
  priority: 'urgent' | 'normal' | 'low';
  payload: Record<string, unknown>;
  scheduledFor?: Date;           // For delayed/scheduled notifications
  createdAt: Date;
}
```

Delivery Guarantees

- At-least-once delivery (idempotency keys prevent duplicates)
- Priority queue for urgent notifications
- Dead letter queue for failed deliveries after 3 retries ...

```
The output is a working first draft. You'll edit it — adding your specific conte
xt, adjusting the tone, removing AI assumptions that don't match reality — but y
ou save hours compared to writing from a blank page.

### Database Schema Design

AI helps you think through data modeling:
```

I need to design a database schema for a multi-tenant project management application. Core concepts:

- Organizations (tenants)
- Users (belong to one or more organizations)
- Projects (belong to an organization)
- Tasks (belong to a project, assigned to users)
- Comments (on tasks, by users)
- Time entries (on tasks, by users)
- Labels (per organization, applied to tasks)

Requirements: - Multi-tenant with row-level security - Soft deletes on all entities - Audit trail (who changed what, when) - Full-text search on tasks and comments

- Efficient queries for dashboards (task counts by status, time entries by date range)

Design the schema in PostgreSQL. Include indexes. Explain your design choices.

````
AI generates a complete schema with thoughtful indexing:

```sql
-- Organizations (tenants)
CREATE TABLE organizations (
    id UUID PRIMARY KEY DEFAULT gen_random_uuid(),
    name TEXT NOT NULL,
    slug TEXT NOT NULL UNIQUE,
    created_at TIMESTAMPTZ NOT NULL DEFAULT now(),
    deleted_at TIMESTAMPTZ,

    -- RLS policy will filter on this
    CONSTRAINT organizations_slug_format CHECK (slug ~ '^[a-z0-9-]+$')
);

-- Users with organization membership
CREATE TABLE users (
    id UUID PRIMARY KEY DEFAULT gen_random_uuid(),
    email TEXT NOT NULL UNIQUE,
    name TEXT NOT NULL,
    created_at TIMESTAMPTZ NOT NULL DEFAULT now(),
    deleted_at TIMESTAMPTZ
);

CREATE TABLE organization_members (
    organization_id UUID NOT NULL REFERENCES organizations(id),
    user_id UUID NOT NULL REFERENCES users(id),
    role TEXT NOT NULL DEFAULT 'member'
        CHECK (role IN ('owner', 'admin', 'member', 'viewer')),
    joined_at TIMESTAMPTZ NOT NULL DEFAULT now(),
    PRIMARY KEY (organization_id, user_id)
);

-- Tasks with full-text search
CREATE TABLE tasks (
    id UUID PRIMARY KEY DEFAULT gen_random_uuid(),
    organization_id UUID NOT NULL REFERENCES organizations(id),
    project_id UUID NOT NULL REFERENCES projects(id),
    title TEXT NOT NULL,
````
````

```sql
    description TEXT,
    status TEXT NOT NULL DEFAULT 'open'
        CHECK (status IN ('open', 'in_progress', 'review', 'done', 'cancelled')),
    assignee_id UUID REFERENCES users(id),
    priority INTEGER NOT NULL DEFAULT 0 CHECK (priority BETWEEN 0 AND 4),
    due_date DATE,
    created_by UUID NOT NULL REFERENCES users(id),
    created_at TIMESTAMPTZ NOT NULL DEFAULT now(),
    updated_at TIMESTAMPTZ NOT NULL DEFAULT now(),
    deleted_at TIMESTAMPTZ,

    -- Full-text search vector
    search_vector TSVECTOR GENERATED ALWAYS AS (
        setweight(to_tsvector('english', coalesce(title, '')), 'A') ||
        setweight(to_tsvector('english', coalesce(description, '')), 'B')
    ) STORED
);

-- Indexes for common dashboard queries
CREATE INDEX idx_tasks_org_status ON tasks(organization_id, status)
    WHERE deleted_at IS NULL;
CREATE INDEX idx_tasks_assignee ON tasks(assignee_id, status)
    WHERE deleted_at IS NULL;
CREATE INDEX idx_tasks_search ON tasks USING GIN(search_vector);
CREATE INDEX idx_tasks_due_date ON tasks(due_date)
    WHERE deleted_at IS NULL AND status NOT IN ('done', 'cancelled');
```

AI explains each choice: why UUIDs over integers (multi-tenant merge safety), why generated search vectors (automatic indexing), why partial indexes on `deleted_at IS NULL` (soft deletes shouldn't bloat indexes).

System Design Interview Practice

Useful for interview prep and for thinking through real systems:

```
Design a URL shortener that handles 100M URLs and 1B redirects/month.

Walk me through:
1. API design
2. Data model
3. URL generation algorithm
4. Read/write path
5. Caching strategy
6. How to handle analytics (click tracking)
```

```
7. How to scale horizontally

Challenge my assumptions at each step.
```

AI will walk through each component, and the back-and-forth surfaces trade-offs you'd miss alone:

- Base62 vs. hash-based short URLs (collision handling)
- Read-through cache vs. cache-aside (consistency vs. performance)
- Synchronous vs. async analytics (latency impact on redirects)
- Sharding strategies (by short URL vs. by creation time)

Architecture Anti-Patterns Detection

```
claude "Analyze the codebase architecture. Identify any of
  these anti-patterns:
  1. God classes (classes that do too many things)
  2. Circular dependencies between modules
  3. Layer violations (routes accessing the database directly)
  4. Inappropriate coupling (services knowing about HTTP concepts)
  5. Missing abstraction layers
  6. Configuration spread across multiple locations

  For each finding, explain why it's problematic and suggest
  how to fix it."
```

The Limits of AI in Architecture

Be aware of where AI guidance on architecture falls short:

AI doesn't know your organizational constraints. It may suggest a microservices architecture when your team of three can't operate it. Always filter AI advice through your team's capabilities.

AI optimizes for what it has seen. It recommends popular patterns. Sometimes your problem needs an unusual solution AI won't suggest because it's rare in training data.

AI doesn't know your future. Architecture is partly about predicting what will change. AI doesn't know your roadmap, your market, or your users' evolving needs.

AI conflates complexity with quality. AI sometimes suggests over-engineered solutions — three services where one would do, abstractions that will never be reused. Simplicity is a feature. Push back on unnecessary complexity.

AI doesn't validate its own diagrams. When AI generates architecture diagrams in Mermaid or PlantUML, the diagram often shows connections the text description doesn't mention, or omits ones that are critical. Always cross-reference the diagram against the prose and your actual system. Diagrams look authoritative but they're just as hallucination-prone as code.

The best approach: use AI to explore the decision space thoroughly, then apply your own judgment to make the call. The decision is yours. The exploration is where AI earns its keep.

Architecture defines the system. Implementation builds it. Chapters 11 through 16 cover domain-specific development with AI assistance — frontend, backend, DevOps, large codebases, git workflows, and API development. Chapter 11 starts with frontend, where AI's ability to generate UI components, CSS, and responsive layouts makes it an indispensable tool for building user interfaces.

Chapter 11: Frontend Development with AI

React, Vue, CSS, Responsive Design, Component Generation

Frontend development is where AI assistance feels most magical. You describe a component in plain English, and AI produces working React or Vue code with proper styling, event handlers, accessibility attributes, and responsive behavior. The gap between "I want a sortable data table with pagination" and having that component in your project has collapsed from hours to minutes.

This chapter covers AI-assisted frontend work across frameworks, from component generation to CSS authoring to responsive design patterns.

Component Generation

React Components

The most common frontend AI task: generating components from descriptions.

```
Create a React component for a user avatar with these requirements:
- Displays user's profile image if available
- Falls back to initials (first letter of first and last name)
  on a colored background
- The background color is deterministic based on the user's name
  (same name always gets same color)
- Sizes: sm (32px), md (40px), lg (64px), xl (96px)
- Shows an online status indicator (green dot) when online prop is true
- Accessible: alt text for image, aria-label for initials
- TypeScript, functional component, Tailwind CSS
- Export the component and its props type
```

AI produces:

```
import { useMemo } from 'react';

interface AvatarProps {
```

```
  name: string;
  imageUrl?: string | null;
  size?: 'sm' | 'md' | 'lg' | 'xl';
  online?: boolean;
  className?: string;
}

const sizeClasses = {
  sm: 'w-8 h-8 text-xs',
  md: 'w-10 h-10 text-sm',
  lg: 'w-16 h-16 text-lg',
  xl: 'w-24 h-24 text-2xl',
} as const;

const statusSizeClasses = {
  sm: 'w-2 h-2',
  md: 'w-2.5 h-2.5',
  lg: 'w-3 h-3',
  xl: 'w-4 h-4',
} as const;

function getInitials(name: string): string {
  const parts = name.trim().split(/\s+/);
  if (parts.length === 0) return '?';
  if (parts.length === 1) return parts[0][0].toUpperCase();
  return (parts[0][0] + parts[parts.length - 1][0]).toUpperCase();
}

function getColorFromName(name: string): string {
  let hash = 0;
  for (const char of name) {
    hash = char.charCodeAt(0) + ((hash << 5) - hash);
  }
  const colors = [
    'bg-red-500', 'bg-blue-500', 'bg-green-500', 'bg-purple-500',
    'bg-yellow-500', 'bg-pink-500', 'bg-indigo-500', 'bg-teal-500',
  ];
  return colors[Math.abs(hash) % colors.length];
}

export function Avatar({
  name,
  imageUrl,
  size = 'md',
  online,
```

```
  className = ''
}: AvatarProps) {
  const initials = useMemo(() => getInitials(name), [name]);
  const bgColor = useMemo(() => getColorFromName(name), [name]);

  return (
    <div className={`relative inline-flex ${className}`}>
      {imageUrl ? (
        <img
          src={imageUrl}
          alt={`${name}'s avatar`}
          className={`${sizeClasses[size]} rounded-full object-cover`}
        />
      ) : (
        <div
          className={`${sizeClasses[size]} ${bgColor} rounded-full flex items-ce
nter justify-center text-white font-medium`}
          aria-label={`${name}'s avatar`}
          role="img"
        >
          {initials}
        </div>
      )}
      {online && (
        <span
          className={`absolute bottom-0 right-0 ${statusSizeClasses[size]} bg-gr
een-400 border-2 border-white rounded-full`}
          aria-label="Online"
        />
      )}
    </div>
  );
}

export type { AvatarProps };
```

This is production-quality code. The accessibility attributes, deterministic color generation, and proper TypeScript types would all take extra time to add manually.

Vue Components

AI handles Vue 3 Composition API equally well:

```
Create a Vue 3 component for a search input with autocomplete.
Use the Composition API with <script setup lang="ts">.
Requirements:
- Debounced search (300ms)
- Shows dropdown with results
- Keyboard navigation (arrow keys, enter to select, escape to close)
- Loading state while fetching
- Empty state message
- Accessible with ARIA attributes
- Emits 'select' event with the chosen item
- Uses Tailwind CSS
```

```
<script setup lang="ts">
import { ref, watch, computed, onMounted, onUnmounted } from 'vue';

interface SearchItem {
  id: string;
  label: string;
  description?: string;
}

interface Props {
  placeholder?: string;
  fetchResults: (query: string) => Promise<SearchItem[]>;
  minChars?: number;
}

const props = withDefaults(defineProps<Props>(), {
  placeholder: 'Search...',
  minChars: 2,
});

const emit = defineEmits<{
  select: [item: SearchItem];
}>();

const query = ref('');
const results = ref<SearchItem[]>([]);
const isOpen = ref(false);
const isLoading = ref(false);
const activeIndex = ref(-1);
const inputRef = ref<HTMLInputElement>();

let debounceTimer: ReturnType<typeof setTimeout>;
```

```
watch(query, (newQuery) => {
  clearTimeout(debounceTimer);
  if (newQuery.length < props.minChars) {
    results.value = [];
    isOpen.value = false;
    return;
  }
  debounceTimer = setTimeout(async () => {
    isLoading.value = true;
    try {
      results.value = await props.fetchResults(newQuery);
      isOpen.value = results.value.length > 0;
      activeIndex.value = -1;
    } finally {
      isLoading.value = false;
    }
  }, 300);
});

function handleKeydown(event: KeyboardEvent) {
  if (!isOpen.value) return;
  switch (event.key) {
    case 'ArrowDown':
      event.preventDefault();
      activeIndex.value = Math.min(activeIndex.value + 1, results.value.length -
1);
      break;
    case 'ArrowUp':
      event.preventDefault();
      activeIndex.value = Math.max(activeIndex.value - 1, 0);
      break;
    case 'Enter':
      if (activeIndex.value >= 0) {
        selectItem(results.value[activeIndex.value]);
      }
      break;
    case 'Escape':
      isOpen.value = false;
      break;
  }
}

function selectItem(item: SearchItem) {
  query.value = item.label;
  isOpen.value = false;
```

```
  emit('select', item);
}

onUnmounted(() => clearTimeout(debounceTimer));
</script>

<template>
  <div class="relative w-full">
    <input
      ref="inputRef"
      v-model="query"
      :placeholder="placeholder"
      type="text"
      role="combobox"
      :aria-expanded="isOpen"
      aria-autocomplete="list"
      aria-controls="search-results"
      class="w-full px-4 py-2 border border-gray-300 rounded-lg focus:ring-2 foc
us:ring-blue-500 focus:border-transparent"
      @keydown="handleKeydown"
    />
    <div v-if="isLoading" class="absolute right-3 top-2.5">
      <svg class="animate-spin h-5 w-5 text-gray-400" viewBox="0 0 24 24">
        <circle class="opacity-25" cx="12" cy="12" r="10" stroke="currentColor"
stroke-width="4" fill="none" />
        <path class="opacity-75" fill="currentColor" d="M4 12a8 8 0 018-8V0C5.37
3 0 0 5.373 0 12h4z" />
      </svg>
    </div>
    <ul
      v-if="isOpen"
      id="search-results"
      role="listbox"
      class="absolute z-10 w-full mt-1 bg-white border border-gray-200 rounded-l
g shadow-lg max-h-60 overflow-auto"
    >
      <li
        v-for="(item, index) in results"
        :key="item.id"
        role="option"
        :aria-selected="index === activeIndex"
        :class="[
          'px-4 py-2 cursor-pointer',
          index === activeIndex ? 'bg-blue-50 text-blue-700' : 'hover:bg-gray-50
'
```

```
        ]"
        @click="selectItem(item)"
        @mouseenter="activeIndex = index"
      >
        <div class="font-medium">{{ item.label }}</div>
        <div v-if="item.description" class="text-sm text-gray-500">
          {{ item.description }}
        </div>
      </li>
    </ul>
  </div>
</template>
```

The Component Generation Workflow in Practice

The examples above show individual prompts. In practice, component generation is iterative — and the iteration is where the real productivity comes from. Here's what a realistic React component workflow looks like, start to finish.

Step 1: Generate the base component.

```
Create a React notification toast component. TypeScript, Tailwind CSS.
Requirements:
- Types: success, error, warning, info
- Auto-dismiss after configurable duration (default 5s)
- Manual dismiss with X button
- Stacks multiple toasts vertically
- Animate in from the top right, animate out on dismiss
- Accessible: role="alert" for errors, role="status" for info
```

AI generates a `Toast` component and a `ToastProvider` with context. The code works, but you notice the `setTimeout` for auto-dismiss isn't cleaned up on unmount.

Step 2: Fix and refine.

```
Good start, but two issues:
1. The setTimeout isn't cleaned up if the component unmounts early
2. I need a useToast() hook that any component can call to show a toast

Also, add a progress bar that shows remaining time before auto-dismiss.
```

AI updates the component with proper cleanup and adds the hook. The progress bar uses a CSS animation synced to the timeout duration.

Step 3: Integrate with your design system.

```
This looks right. Now adapt it to match our existing design system:
- Use our color tokens from tailwind.config.ts (colors.success, colors.danger, e
tc.)
- Match the border-radius we use elsewhere (rounded-xl)
- Use our standard shadow (shadow-card)
- The X button should use our IconButton component from @/components/ui
```

This step is where AI saves the most time. Adapting a generic component to your design system involves dozens of small decisions about spacing, colors, and typography. Hand AI your conventions and it applies them consistently in one pass.

Step 4: Add tests.

```
Write tests for the toast system using React Testing Library.
Test that toasts appear, auto-dismiss after the timeout,
can be manually dismissed, and that the useToast hook works
from a child component.
```

The whole workflow — from "I need a toast component" to tested, design-system-integrated code — takes about 20 minutes. Without AI, it's easily a two-hour task, because toast systems involve surprising amounts of coordination between the provider, the hook, the animations, and the stacking logic.

What AI gets wrong in this workflow: It tends to over-scope the initial generation. Ask for a toast component and AI might give you a toast component, a notification center, a persistence layer, and a preferences panel. Be specific about what you need now versus what you might need later.

CSS and Styling with AI

AI is remarkably good at CSS, which is welcome news because CSS is remarkably good at surprising you.

Tailwind CSS Generation

```
Convert this design description to Tailwind CSS classes for a card component:
- White background with subtle shadow
- Rounded corners (8px)
- Padding: 24px
- Title: 18px bold, dark gray
```

```
- Subtitle: 14px regular, medium gray
- Bottom border separating header from content
- Hover: slightly elevated shadow
- Responsive: full width on mobile, 1/3 width on desktop
```

AI produces:

```
<div class="w-full md:w-1/3 bg-white rounded-lg shadow-sm hover:shadow-md
            transition-shadow duration-200 p-6">
  <div class="border-b border-gray-100 pb-4 mb-4">
    <h3 class="text-lg font-bold text-gray-800">Card Title</h3>
    <p class="text-sm text-gray-500 mt-1">Card subtitle</p>
  </div>
  <div>
    <!-- Card content -->
  </div>
</div>
```

Complex CSS Layouts

For layouts that would otherwise eat 30 minutes of CSS wrestling:

```
Create a CSS Grid layout for a dashboard:
- Sidebar: 240px fixed width, full height, collapsible on mobile
- Top navbar: 64px height, spans full width except sidebar
- Main content: fills remaining space, scrollable
- Footer: sticks to bottom, only visible when scrolled to end
- On mobile (<768px): sidebar becomes a slide-out drawer

Use plain CSS with CSS Grid and custom properties.
No framework dependencies.
```

```
:root {
  --sidebar-width: 240px;
  --navbar-height: 64px;
  --sidebar-transition: transform 0.3s ease;
}

.dashboard {
  display: grid;
  grid-template-areas:
    "sidebar navbar"
    "sidebar content"
    "sidebar footer";
  grid-template-columns: var(--sidebar-width) 1fr;
```

```css
  grid-template-rows: var(--navbar-height) 1fr auto;
  min-height: 100vh;
}

.sidebar {
  grid-area: sidebar;
  background: #1a1a2e;
  color: white;
  overflow-y: auto;
  position: sticky;
  top: 0;
  height: 100vh;
}

.navbar {
  grid-area: navbar;
  background: white;
  border-bottom: 1px solid #e5e7eb;
  display: flex;
  align-items: center;
  padding: 0 24px;
  position: sticky;
  top: 0;
  z-index: 10;
}

.content {
  grid-area: content;
  padding: 24px;
  overflow-y: auto;
}

.footer {
  grid-area: footer;
  padding: 16px 24px;
  border-top: 1px solid #e5e7eb;
  background: #f9fafb;
}

/* Mobile: sidebar becomes drawer */
@media (max-width: 767px) {
  .dashboard {
    grid-template-areas:
      "navbar"
      "content"
```

```
    "footer";
  grid-template-columns: 1fr;
}

.sidebar {
  position: fixed;
  top: 0;
  left: 0;
  height: 100vh;
  width: var(--sidebar-width);
  transform: translateX(-100%);
  transition: var(--sidebar-transition);
  z-index: 50;
}

.sidebar.open {
  transform: translateX(0);
}

.sidebar-overlay {
  display: none;
  position: fixed;
  inset: 0;
  background: rgba(0, 0, 0, 0.5);
  z-index: 40;
}

.sidebar.open ~ .sidebar-overlay {
  display: block;
}
}
```

Responsive Design

AI handles responsive design well because the patterns are well-established:

```
Make this product card responsive:
- Mobile (< 640px): Full width, stacked layout (image on top, details below)
- Tablet (640-1024px): Horizontal layout (image left, details right)
- Desktop (> 1024px): Same as tablet but with additional info panel on right

Use Tailwind CSS. Include the image, product name, price,
rating stars, and "Add to Cart" button.
```

Accessibility

AI generates proper accessibility attributes, but you need to ask:

```
Audit this component for accessibility issues. Check:
- ARIA attributes
- Keyboard navigation
- Screen reader announcements
- Focus management
- Color contrast
- Motion preferences (prefers-reduced-motion)

Suggest specific fixes for each issue found.
```

Common AI-generated accessibility improvements:

```
// Before: inaccessible modal
<div className="modal" onClick={onClose}>
  <div className="modal-content">{children}</div>
</div>

// After: accessible modal
<div
  className="modal"
  role="dialog"
  aria-modal="true"
  aria-labelledby="modal-title"
  onClick={onClose}
  onKeyDown={(e) => e.key === 'Escape' && onClose()}
>
  <div
    className="modal-content"
    onClick={(e) => e.stopPropagation()}
    ref={contentRef}
  >
    <h2 id="modal-title">{title}</h2>
    {children}
    <button onClick={onClose} aria-label="Close dialog">
      <XIcon />
    </button>
  </div>
</div>
```

Form Handling

Forms are boilerplate-heavy, and AI handles them well:

```
Create a multi-step registration form with React Hook Form and Zod:
Step 1: Personal info (name, email, phone)
Step 2: Account setup (username, password, confirm password)
Step 3: Preferences (notification settings, timezone, language)
Step 4: Review and submit

Requirements:
- Validation at each step before proceeding
- Progress indicator
- Back button to return to previous steps (preserving data)
- Submit button only on final step
- Show validation errors inline
- Accessible form labels and error messages
```

AI generates the complete multi-step form with proper validation, state management, and UX. AI saves substantial time here because multi-step forms involve significant coordination code.

Frontend Testing with AI

```
Write tests for the SearchAutocomplete component using
React Testing Library and Vitest:
- Renders without crashing
- Shows results after typing (with debounce)
- Keyboard navigation works
- Selecting an item calls onSelect
- Escape key closes dropdown
- Shows loading state during fetch
- Shows empty state when no results

Use userEvent for interactions. Mock the fetchResults function.
```

```
import { render, screen, waitFor } from '@testing-library/react';
import userEvent from '@testing-library/user-event';
import { vi, describe, it, expect } from 'vitest';
import { SearchAutocomplete } from './SearchAutocomplete';

const mockResults = [
  { id: '1', label: 'React', description: 'A JavaScript library' },
  { id: '2', label: 'Redux', description: 'State management' },
```

```
];

const mockFetch = vi.fn().mockResolvedValue(mockResults);

describe('SearchAutocomplete', () => {
  it('renders input with placeholder', () => {
    render(<SearchAutocomplete fetchResults={mockFetch} placeholder="Search..."
/>);
    expect(screen.getByPlaceholderText('Search...')).toBeInTheDocument();
  });

  it('shows results after typing with debounce', async () => {
    const user = userEvent.setup();
    render(<SearchAutocomplete fetchResults={mockFetch} />);

    await user.type(screen.getByRole('combobox'), 'rea');

    await waitFor(() => {
      expect(screen.getByText('React')).toBeInTheDocument();
      expect(screen.getByText('Redux')).toBeInTheDocument();
    });
  });

  it('navigates results with arrow keys', async () => {
    const user = userEvent.setup();
    render(<SearchAutocomplete fetchResults={mockFetch} />);

    await user.type(screen.getByRole('combobox'), 'rea');
    await waitFor(() => screen.getByText('React'));

    await user.keyboard('{ArrowDown}');
    expect(screen.getByText('React').closest('[role="option"]'))
      .toHaveAttribute('aria-selected', 'true');
  });

  it('closes dropdown on Escape', async () => {
    const user = userEvent.setup();
    render(<SearchAutocomplete fetchResults={mockFetch} />);

    await user.type(screen.getByRole('combobox'), 'rea');
    await waitFor(() => screen.getByText('React'));

    await user.keyboard('{Escape}');
    expect(screen.queryByRole('listbox')).not.toBeInTheDocument();
```

```
  });
});
```

What AI Gets Wrong in Frontend

- **Outdated React patterns:** Class components, `componentDidMount`, `this.setState` — tell AI to use hooks
- **Inline styles instead of your CSS framework:** Specify Tailwind, CSS modules, or styled-components
- **Missing error boundaries:** AI rarely adds error boundaries unless asked
- **Ignoring loading states:** AI generates the happy-path UI and forgets loading, error, and empty states
- **Over-engineering:** AI may add Redux when local state would suffice, or pull in a library for something CSS can do natively

Always specify your stack and preferences in the prompt. The more context AI has about your project's frontend conventions, the less editing you do.

Chapter 12: Backend Development with AI

APIs, Databases, Server Logic, Microservices

Backend is where AI assistance shifts from "nice to have" to genuine productivity multiplier. Backend code follows well-worn patterns — request handling, validation, business logic, data access, response formatting — and AI knows them cold.

This chapter covers AI-assisted backend work across languages and frameworks: building APIs, designing data access layers, and structuring services.

API Endpoint Generation

Express/Node.js

```
Create a complete Express route module for managing products.
Include:
- GET /products (list with pagination, sorting, filtering)
- GET /products/:id (single product)
- POST /products (create, requires admin role)
- PUT /products/:id (update, requires admin role)
- DELETE /products/:id (soft delete, requires admin role)

Use TypeScript, Zod validation, Prisma ORM, and our standard
middleware patterns:
- asyncHandler wrapper for error handling
- requireAuth middleware
- requireRole('admin') middleware
- validateBody(schema) middleware for request validation
- Standard response envelope: { data, error, meta }
```

AI generates a complete, production-ready route module:

```
import { Router } from 'express';
import { z } from 'zod';
import { prisma } from '../lib/prisma';
import { asyncHandler } from '../middleware/async-handler';
```

```
import { requireAuth } from '../middleware/auth';
import { requireRole } from '../middleware/roles';
import { validateBody, validateQuery } from '../middleware/validate';
import { NotFoundError } from '../errors';

const router = Router();

// Schemas
const listQuerySchema = z.object({
  page: z.coerce.number().int().positive().default(1),
  limit: z.coerce.number().int().min(1).max(100).default(20),
  sort: z.enum(['name', 'price', 'createdAt']).default('createdAt'),
  order: z.enum(['asc', 'desc']).default('desc'),
  category: z.string().optional(),
  search: z.string().optional(),
});

const createProductSchema = z.object({
  name: z.string().min(1).max(255),
  description: z.string().max(2000).optional(),
  price: z.string().regex(/^\d+\.\d{2}$/, 'Price must be in format 0.00'),
  category: z.string().min(1),
  sku: z.string().min(1).max(50),
  stock: z.number().int().min(0).default(0),
});

const updateProductSchema = createProductSchema.partial();

// GET /products
router.get(
  '/',
  validateQuery(listQuerySchema),
  asyncHandler(async (req, res) => {
    const { page, limit, sort, order, category, search } = req.validatedQuery;

    const where = {
      deletedAt: null,
      ...(category && { category }),
      ...(search && {
        OR: [
          { name: { contains: search, mode: 'insensitive' as const } },
          { description: { contains: search, mode: 'insensitive' as const } },
        ],
      }),
    };
```

```
    const [products, total] = await Promise.all([
      prisma.product.findMany({
        where,
        orderBy: { [sort]: order },
        skip: (page - 1) * limit,
        take: limit,
      }),
      prisma.product.count({ where }),
    ]);

    res.json({
      data: products,
      error: null,
      meta: {
        page,
        limit,
        total,
        totalPages: Math.ceil(total / limit),
      },
    });
  })
);

// GET /products/:id
router.get(
  '/:id',
  asyncHandler(async (req, res) => {
    const product = await prisma.product.findFirst({
      where: { id: req.params.id, deletedAt: null },
    });

    if (!product) {
      throw new NotFoundError('Product not found');
    }

    res.json({ data: product, error: null, meta: {} });
  })
);

// POST /products
router.post(
  '/',
  requireAuth,
  requireRole('admin'),
```

```
  validateBody(createProductSchema),
  asyncHandler(async (req, res) => {
    const product = await prisma.product.create({
      data: req.validatedBody,
    });

    res.status(201).json({ data: product, error: null, meta: {} });
  })
);

// PUT /products/:id
router.put(
  '/:id',
  requireAuth,
  requireRole('admin'),
  validateBody(updateProductSchema),
  asyncHandler(async (req, res) => {
    const existing = await prisma.product.findFirst({
      where: { id: req.params.id, deletedAt: null },
    });

    if (!existing) {
      throw new NotFoundError('Product not found');
    }

    const product = await prisma.product.update({
      where: { id: req.params.id },
      data: { ...req.validatedBody, updatedAt: new Date() },
    });

    res.json({ data: product, error: null, meta: {} });
  })
);

// DELETE /products/:id (soft delete)
router.delete(
  '/:id',
  requireAuth,
  requireRole('admin'),
  asyncHandler(async (req, res) => {
    const existing = await prisma.product.findFirst({
      where: { id: req.params.id, deletedAt: null },
    });

    if (!existing) {
```

```
      throw new NotFoundError('Product not found');
    }

    await prisma.product.update({
      where: { id: req.params.id },
      data: { deletedAt: new Date() },
    });

    res.status(204).end();
  })
);

export { router as productRoutes };
```

FastAPI/Python

```python
## Prompt: "Create a FastAPI router for user management with
## SQLAlchemy async, Pydantic v2 models, and dependency injection"

from fastapi import APIRouter, Depends, HTTPException, Query
from sqlalchemy.ext.asyncio import AsyncSession
from pydantic import BaseModel, EmailStr, Field
from datetime import datetime
from uuid import UUID

from app.database import get_session
from app.models import User
from app.auth import get_current_user, require_admin

router = APIRouter(prefix="/users", tags=["users"])

class UserCreate(BaseModel):
    email: EmailStr
    name: str = Field(min_length=1, max_length=100)
    role: str = Field(default="member", pattern="^(admin|member|viewer)$")

class UserResponse(BaseModel):
    id: UUID
    email: str
    name: str
    role: str
    created_at: datetime

    model_config = {"from_attributes": True}
```

```python
class UserListResponse(BaseModel):
    data: list[UserResponse]
    total: int
    page: int
    limit: int

@router.get("", response_model=UserListResponse)
async def list_users(
    page: int = Query(default=1, ge=1),
    limit: int = Query(default=20, ge=1, le=100),
    search: str | None = Query(default=None),
    session: AsyncSession = Depends(get_session),
    current_user: User = Depends(get_current_user),
):
    """List all users with pagination and optional search."""
    query = select(User).where(User.deleted_at.is_(None))

    if search:
        query = query.where(
            or_(
                User.name.ilike(f"%{search}%"),
                User.email.ilike(f"%{search}%"),
            )
        )

    total = await session.scalar(
        select(func.count()).select_from(query.subquery())
    )

    users = await session.scalars(
        query.offset((page - 1) * limit).limit(limit)
    )

    return UserListResponse(
        data=[UserResponse.model_validate(u) for u in users],
        total=total,
        page=page,
        limit=limit,
    )

@router.post("", response_model=UserResponse, status_code=201)
async def create_user(
    body: UserCreate,
    session: AsyncSession = Depends(get_session),
    current_user: User = Depends(require_admin),
```

```
):
    """Create a new user. Requires admin role."""
    existing = await session.scalar(
        select(User).where(User.email == body.email)
    )
    if existing:
        raise HTTPException(
            status_code=409,
            detail="A user with this email already exists"
        )

    user = User(**body.model_dump())
    session.add(user)
    await session.commit()
    await session.refresh(user)

    return UserResponse.model_validate(user)
```

Go/Gin

```go
// Prompt: "Create a Go handler for order management using Gin,
// pgx for database, and structured error handling"

package handlers

import (
    "context"
    "net/http"
    "strconv"

    "github.com/gin-gonic/gin"
    "github.com/jackc/pgx/v5/pgxpool"
)

type OrderHandler struct {
    db *pgxpool.Pool
}

func NewOrderHandler(db *pgxpool.Pool) *OrderHandler {
    return &OrderHandler{db: db}
}

type CreateOrderRequest struct {
    Items []OrderItemRequest `json:"items" binding:"required,min=1"`
}
```

```
type OrderItemRequest struct {
    ProductID string `json:"product_id" binding:"required"`
    Quantity  int    `json:"quantity" binding:"required,min=1"`
}

func (h *OrderHandler) CreateOrder(c *gin.Context) {
    var req CreateOrderRequest
    if err := c.ShouldBindJSON(&req); err != nil {
        c.JSON(http.StatusBadRequest, gin.H{
            "error": gin.H{"code": "VALIDATION_ERROR", "message": err.Error()},
        })
        return
    }

    userID, _ := c.Get("userID")

    tx, err := h.db.Begin(c.Request.Context())
    if err != nil {
        c.JSON(http.StatusInternalServerError, gin.H{
            "error": gin.H{"code": "INTERNAL_ERROR", "message": "Failed to start
 transaction"},
        })
        return
    }
    defer tx.Rollback(c.Request.Context())

    // Create order
    var orderID string
    err = tx.QueryRow(c.Request.Context(),
        `INSERT INTO orders (user_id, status) VALUES ($1, 'pending') RETURNING i
d`,
        userID,
    ).Scan(&orderID)
    if err != nil {
        c.JSON(http.StatusInternalServerError, gin.H{
            "error": gin.H{"code": "INTERNAL_ERROR", "message": "Failed to creat
e order"},
        })
        return
    }

    // Add items with inventory check
    for _, item := range req.Items {
        var stock int
```

```
        err = tx.QueryRow(c.Request.Context(),
            `SELECT stock FROM products WHERE id = $1 FOR UPDATE`,
            item.ProductID,
        ).Scan(&stock)
        if err != nil {
            c.JSON(http.StatusNotFound, gin.H{
                "error": gin.H{"code": "PRODUCT_NOT_FOUND", "message": "Product
not found"},
            })
            return
        }
        if stock < item.Quantity {
            c.JSON(http.StatusConflict, gin.H{
                "error": gin.H{"code": "INSUFFICIENT_STOCK", "message": "Not eno
ugh stock"},
            })
            return
        }

        // Reserve stock and add order item
        _, err = tx.Exec(c.Request.Context(),
            `UPDATE products SET stock = stock - $1 WHERE id = $2`,
            item.Quantity, item.ProductID,
        )
        if err != nil {
            c.JSON(http.StatusInternalServerError, gin.H{
                "error": gin.H{"code": "INTERNAL_ERROR", "message": "Failed to u
pdate stock"},
            })
            return
        }

        _, err = tx.Exec(c.Request.Context(),
            `INSERT INTO order_items (order_id, product_id, quantity) VALUES ($1,
 $2, $3)`,
            orderID, item.ProductID, item.Quantity,
        )
        if err != nil {
            c.JSON(http.StatusInternalServerError, gin.H{
                "error": gin.H{"code": "INTERNAL_ERROR", "message": "Failed to a
dd order item"},
            })
            return
        }
    }
```

```
    if err := tx.Commit(c.Request.Context()); err != nil {
        c.JSON(http.StatusInternalServerError, gin.H{
            "error": gin.H{"code": "INTERNAL_ERROR", "message": "Failed to commi
t order"},
        })
        return
    }

    c.JSON(http.StatusCreated, gin.H{
        "data": gin.H{"id": orderID, "status": "pending"},
    })
}

func (h *OrderHandler) RegisterRoutes(r *gin.RouterGroup) {
    orders := r.Group("/orders")
    orders.POST("", h.CreateOrder)
    orders.GET("", h.ListOrders)
    orders.GET("/:id", h.GetOrder)
}
```

Before and After: Adding an Endpoint to an Existing API

The examples above generate a new route module from scratch. In practice, you more often add endpoints to an existing codebase. This is where AI shines — it matches your patterns if you show it what those patterns look like.

The scenario: you have an Express API with an established pattern for route handlers, and you need a new endpoint for user activity tracking.

Before (what you have):

```
// Your existing pattern in src/routes/users.ts
router.get(
  '/:id/profile',
  requireAuth,
  asyncHandler(async (req, res) => {
    const profile = await userService.getProfile(req.params.id);
    if (!profile) throw new NotFoundError('User profile not found');
    res.json({ data: profile, error: null, meta: {} });
  })
);
```

Your prompt:

```
Here's our existing route pattern for GET /:id/profile [paste above].

Now add these endpoints following the same pattern:
- GET /users/:id/activity — returns paginated activity feed
- POST /users/:id/activity — logs a new activity event
- The activity feed supports filtering by type (login, purchase, settings_change)
- The POST endpoint should validate the activity type against an enum
- Use our existing pagination pattern (page/limit query params)
```

After (what AI generates):

```
const activityTypeSchema = z.enum([
  'login', 'purchase', 'settings_change', 'password_reset', 'export'
]);

const activityQuerySchema = z.object({
  page: z.coerce.number().int().positive().default(1),
  limit: z.coerce.number().int().min(1).max(100).default(20),
  type: activityTypeSchema.optional(),
});

const createActivitySchema = z.object({
  type: activityTypeSchema,
  metadata: z.record(z.unknown()).optional(),
});

router.get(
  '/:id/activity',
  requireAuth,
  validateQuery(activityQuerySchema),
  asyncHandler(async (req, res) => {
    const { page, limit, type } = req.validatedQuery;
    const { activities, total } = await activityService.listByUser(
      req.params.id, { page, limit, type }
    );
    res.json({
      data: activities,
      error: null,
      meta: { page, limit, total, totalPages: Math.ceil(total / limit) },
    });
  })
);

router.post(
  '/:id/activity',
```

```
  requireAuth,
  validateBody(createActivitySchema),
  asyncHandler(async (req, res) => {
    const activity = await activityService.create({
      userId: req.params.id,
      ...req.validatedBody,
      ip: req.ip,
    });
    res.status(201).json({ data: activity, error: null, meta: {} });
  })
);
```

Notice what AI did well: it matched your response envelope (`{ data, error, meta }`), used your middleware chain (`requireAuth`, `validateQuery`, `asyncHandler`), and followed your naming conventions. It even added the IP address to the activity log — a nice touch.

Notice what you need to fix: the POST endpoint never verifies that the authenticated user matches `:id` or holds admin role. AI handled authentication (is the user logged in?) but skipped authorization (can this user log activity for this specific user?). That's one of the most consistent blind spots in AI-generated backend code.

Middleware and Cross-Cutting Concerns

AI generates middleware well because middleware patterns are standardized:

```
Create Express middleware for:
1. Request logging (method, path, status, duration, request ID)
2. Error handling (catch all errors, format as standard response)
3. Rate limiting per API key (Redis-backed, configurable)
4. Request validation (Zod schema validation for body/query/params)
```

```
// Request logging middleware
import { randomUUID } from 'crypto';
import { Request, Response, NextFunction } from 'express';
import { logger } from '../lib/logger';

export function requestLogger(req: Request, res: Response, next: NextFunction) {
  const requestId = randomUUID();
  const start = performance.now();

  req.requestId = requestId;
```

```
  res.setHeader('X-Request-ID', requestId);

  res.on('finish', () => {
    const duration = Math.round(performance.now() - start);
    logger.info({
      method: req.method,
      path: req.path,
      status: res.statusCode,
      duration: `${duration}ms`,
      requestId,
      userAgent: req.get('user-agent'),
      ip: req.ip,
    });
  });

  next();
}

// Error handling middleware
export function errorHandler(
  err: Error,
  req: Request,
  res: Response,
  _next: NextFunction
) {
  if (err instanceof AppError) {
    res.status(err.statusCode).json({
      data: null,
      error: { code: err.code, message: err.message },
      meta: { requestId: req.requestId },
    });
    return;
  }

  logger.error({
    error: err.message,
    stack: err.stack,
    requestId: req.requestId,
    path: req.path,
  });

  res.status(500).json({
    data: null,
    error: { code: 'INTERNAL_ERROR', message: 'An unexpected error occurred' },
    meta: { requestId: req.requestId },
```

```
  });
}
```

Background Jobs and Workers

```
Create a background job processing system using BullMQ (Redis-based).
Jobs:
1. SendEmail – sends transactional emails via AWS SES
2. GenerateReport – creates PDF reports, stores in S3
3. SyncInventory – syncs inventory with warehouse API

Requirements:
- Retry failed jobs (3 attempts with exponential backoff)
- Dead letter queue for permanently failed jobs
- Job progress tracking
- Graceful shutdown
- TypeScript
```

AI produces complete worker definitions with error handling, retries, and monitoring — all patterns it has seen thousands of times.

Service Layer Patterns

```
Create a service layer for our e-commerce application. Services:
- OrderService: create, cancel, refund orders
- InventoryService: check, reserve, release stock
- PaymentService: charge, refund, get payment status
- NotificationService: send order confirmation, shipping updates

Requirements:
- Services use dependency injection (constructor parameters)
- Services communicate through interfaces (not concrete implementations)
- Transactions span multiple services using a UnitOfWork pattern
- Each service has its own error types
```

AI understands these enterprise patterns and generates clean, well-structured service layers.

Database Access Patterns

AI generates data access code well when you tell it which pattern you use. Without guidance, it mixes patterns freely — raw SQL in one handler, ORM in the next, a query builder in the third.

```
Our codebase uses the repository pattern for all database access.
Here's an example of our UserRepository:
[paste example]

Generate a ProductRepository following the same pattern. Include:
- findById, findMany (with filters and pagination)
- create, update, softDelete
- Reuse our base repository class from src/repositories/base.ts
- All methods return domain types, not database rows
```

Show AI your existing patterns and tell it to follow them, and the output is far more consistent than when you let AI choose its own approach. The principle holds everywhere, but it matters most in backend code, where inconsistency creates maintenance headaches across dozens of route handlers.

What AI Gets Wrong in Backend

Transaction boundaries: AI sometimes puts database operations outside transactions that should contain them, or draws transactions too broadly (holding locks unnecessarily).

Connection management: AI forgets to close connections, pool properly, or handle connection failures. Always verify the connection lifecycle.

Concurrency: Backend code runs concurrently. AI produces code that works in single-threaded testing but fails under concurrent load — shared mutable state, race conditions.

Secrets in code: AI regularly drops API keys, connection strings, and passwords directly into source. Always grep for hardcoded secrets.

Missing input validation at the boundary: AI sometimes validates at the wrong layer — buried inside business logic instead of at the API boundary, or not at all.

Review AI-generated backend code for these specific issues before deployment.

From Server Logic to Server Substrate

Backend code runs on infrastructure. Chapter 13 covers AI-assisted DevOps — CI/CD pipelines, Docker containers, Kubernetes manifests, and infrastructure as code. AI saves enormous time here because infrastructure config is intricate, repetitive, and unforgiving of errors.

Chapter 13: DevOps and Infrastructure

CI/CD, Docker, Kubernetes, IaC, Monitoring

Infrastructure configuration is the domain where AI saves you the most time and prevents the most costly errors. DevOps work is pattern-heavy, syntax-sensitive, and unforgiving — a misindented YAML file can bring down your deployment pipeline, and a misconfigured security group can expose your database to the internet.

AI knows these patterns cold. It has seen every variation of every CI/CD pipeline, Dockerfile, Kubernetes manifest, and Terraform configuration. Your job is to describe what you need precisely, then verify the output against your security requirements.

Docker

Generating Dockerfiles

```
Create a production Dockerfile for a Node.js TypeScript application:
- Multi-stage build (build stage and runtime stage)
- Use node:20-alpine as base
- Install only production dependencies in runtime
- Non-root user
- Health check endpoint at /health
- Proper signal handling (SIGTERM)
- .dockerignore file
- Optimized layer caching (copy package files first)
```

```
## Build stage
FROM node:20-alpine AS builder

WORKDIR /app

## Copy dependency files first (layer caching)
COPY package.json package-lock.json ./
RUN npm ci

## Copy source and build
COPY tsconfig.json ./
```

```
COPY src/ ./src/
RUN npm run build

## Prune dev dependencies
RUN npm prune --production

## Runtime stage
FROM node:20-alpine AS runtime

## Security: non-root user
RUN addgroup -g 1001 -S appgroup && \
    adduser -S appuser -u 1001 -G appgroup

WORKDIR /app

## Copy only what we need
COPY --from=builder --chown=appuser:appgroup /app/dist ./dist
COPY --from=builder --chown=appuser:appgroup /app/node_modules ./node_modules
COPY --from=builder --chown=appuser:appgroup /app/package.json ./

USER appuser

EXPOSE 3000

HEALTHCHECK --interval=30s --timeout=3s --retries=3 \
  CMD wget -qO- http://localhost:3000/health || exit 1

## Use tini for proper signal handling
RUN apk add --no-cache tini
ENTRYPOINT ["/sbin/tini", "--"]

CMD ["node", "dist/index.js"]

## .dockerignore
node_modules
npm-debug.log
dist
.git
.github
.env
.env.*
*.md
tests/
coverage/
.vscode/
```

Docker Compose for Development

```
Create a docker-compose.yml for local development:
- Node.js app with hot reload (mount source as volume)
- PostgreSQL 16 with persistent volume
- Redis 7 for caching and job queues
- Adminer for database management UI
- Mailpit for local email testing
- All services on a shared network
- Environment variables from .env file
```

```yaml
services:
  app:
    build:
      context: .
      dockerfile: Dockerfile.dev
    ports:
      - "3000:3000"
      - "9229:9229"  # Node.js debugger
    volumes:
      - ./src:/app/src
      - ./tests:/app/tests
    env_file: .env
    environment:
      - DATABASE_URL=postgresql://postgres:postgres@db:5432/app_dev
      - REDIS_URL=redis://redis:6379
      - SMTP_HOST=mailpit
      - SMTP_PORT=1025
    depends_on:
      db:
        condition: service_healthy
      redis:
        condition: service_healthy
    networks:
      - app-network

  db:
    image: postgres:16-alpine
    ports:
      - "5432:5432"
    environment:
      POSTGRES_USER: postgres
      POSTGRES_PASSWORD: postgres
      POSTGRES_DB: app_dev
    volumes:
```

```yaml
      - postgres_data:/var/lib/postgresql/data
      - ./scripts/init-db.sql:/docker-entrypoint-initdb.d/init.sql
    healthcheck:
      test: ["CMD-SHELL", "pg_isready -U postgres"]
      interval: 5s
      timeout: 5s
      retries: 5
    networks:
      - app-network

  redis:
    image: redis:7-alpine
    ports:
      - "6379:6379"
    healthcheck:
      test: ["CMD", "redis-cli", "ping"]
      interval: 5s
      timeout: 5s
      retries: 5
    networks:
      - app-network

  adminer:
    image: adminer
    ports:
      - "8080:8080"
    networks:
      - app-network

  mailpit:
    image: axllent/mailpit
    ports:
      - "8025:8025"  # Web UI
      - "1025:1025"  # SMTP
    networks:
      - app-network

volumes:
  postgres_data:

networks:
  app-network:
    driver: bridge
```

CI/CD Pipelines

GitHub Actions

```
Create a GitHub Actions CI/CD pipeline for a Node.js TypeScript
application with:
- Lint, type-check, and test on every PR
- Build and push Docker image on merge to main
- Deploy to staging automatically
- Deploy to production on manual approval
- Cache node_modules between runs
- Run tests in parallel (unit and integration separately)
- Notify Slack on failure
- Use OIDC for AWS authentication (no long-lived credentials)
```

```yaml
name: CI/CD Pipeline

on:
  push:
    branches: [main]
  pull_request:
    branches: [main]

permissions:
  id-token: write
  contents: read

env:
  NODE_VERSION: '20'
  ECR_REGISTRY: ${{ secrets.AWS_ACCOUNT_ID }}.dkr.ecr.us-east-1.amazonaws.com
  ECR_REPOSITORY: myapp

jobs:
  lint-and-typecheck:
    runs-on: ubuntu-latest
    steps:
      - uses: actions/checkout@v4

      - uses: actions/setup-node@v4
        with:
          node-version: ${{ env.NODE_VERSION }}
          cache: 'npm'

      - run: npm ci
```

```yaml
      - run: npm run lint
      - run: npm run typecheck

  unit-tests:
    runs-on: ubuntu-latest
    steps:
      - uses: actions/checkout@v4

      - uses: actions/setup-node@v4
        with:
          node-version: ${{ env.NODE_VERSION }}
          cache: 'npm'

      - run: npm ci
      - run: npm run test:unit -- --coverage

      - uses: actions/upload-artifact@v4
        with:
          name: coverage-report
          path: coverage/

  integration-tests:
    runs-on: ubuntu-latest
    services:
      postgres:
        image: postgres:16-alpine
        env:
          POSTGRES_USER: test
          POSTGRES_PASSWORD: test
          POSTGRES_DB: app_test
        ports:
          - 5432:5432
        options: >-
          --health-cmd pg_isready
          --health-interval 10s
          --health-timeout 5s
          --health-retries 5
      redis:
        image: redis:7-alpine
        ports:
          - 6379:6379
    steps:
      - uses: actions/checkout@v4

      - uses: actions/setup-node@v4
```

```yaml
        with:
          node-version: ${{ env.NODE_VERSION }}
          cache: 'npm'

      - run: npm ci
      - run: npm run db:migrate
        env:
          DATABASE_URL: postgresql://test:test@localhost:5432/app_test
      - run: npm run test:integration
        env:
          DATABASE_URL: postgresql://test:test@localhost:5432/app_test
          REDIS_URL: redis://localhost:6379

  build-and-push:
    needs: [lint-and-typecheck, unit-tests, integration-tests]
    if: github.ref == 'refs/heads/main'
    runs-on: ubuntu-latest
    outputs:
      image-tag: ${{ steps.meta.outputs.tags }}
    steps:
      - uses: actions/checkout@v4

      - uses: aws-actions/configure-aws-credentials@v4
        with:
          role-to-assume: ${{ secrets.AWS_DEPLOY_ROLE_ARN }}
          aws-region: us-east-1

      - uses: aws-actions/amazon-ecr-login@v2

      - name: Build and push
        uses: docker/build-push-action@v5
        with:
          push: true
          tags: |
            ${{ env.ECR_REGISTRY }}/${{ env.ECR_REPOSITORY }}:${{ github.sha }}
            ${{ env.ECR_REGISTRY }}/${{ env.ECR_REPOSITORY }}:latest
          cache-from: type=gha
          cache-to: type=gha,mode=max

  deploy-staging:
    needs: build-and-push
    if: github.ref == 'refs/heads/main'
    runs-on: ubuntu-latest
    environment: staging
    steps:
```

```yaml
      - uses: aws-actions/configure-aws-credentials@v4
        with:
          role-to-assume: ${{ secrets.AWS_DEPLOY_ROLE_ARN }}
          aws-region: us-east-1

      - name: Deploy to ECS
        run: |
          aws ecs update-service \
            --cluster staging \
            --service myapp \
            --force-new-deployment

  deploy-production:
    needs: deploy-staging
    runs-on: ubuntu-latest
    environment: production  # Requires manual approval
    steps:
      - uses: aws-actions/configure-aws-credentials@v4
        with:
          role-to-assume: ${{ secrets.AWS_DEPLOY_ROLE_ARN }}
          aws-region: us-east-1

      - name: Deploy to ECS
        run: |
          aws ecs update-service \
            --cluster production \
            --service myapp \
            --force-new-deployment

  notify-failure:
    needs: [lint-and-typecheck, unit-tests, integration-tests, build-and-push]
    if: failure()
    runs-on: ubuntu-latest
    steps:
      - uses: slackapi/slack-github-action@v1
        with:
          payload: |
            {
              "text": "CI failed on ${{ github.repository }}: ${{ github.event.h
ead_commit.message }}"
            }
        env:
          SLACK_WEBHOOK_URL: ${{ secrets.SLACK_WEBHOOK_URL }}
```

Kubernetes

```
Create Kubernetes manifests for deploying a web application:
- Deployment with 3 replicas, rolling updates
- Service (ClusterIP)
- Ingress with TLS (cert-manager)
- HorizontalPodAutoscaler (scale 3-10 based on CPU)
- ConfigMap for application config
- Secret for sensitive values
- PodDisruptionBudget
- Resource limits and requests
- Liveness and readiness probes
```

AI generates these manifests with best practices — resource requests/limits, security contexts, anti-affinity rules, and proper health checks.

Infrastructure as Code

Terraform

```
Create Terraform configuration for an AWS ECS Fargate service:
- VPC with public and private subnets across 2 AZs
- ALB with HTTPS listener and certificate
- ECS cluster with Fargate capacity provider
- ECS service with task definition
- RDS PostgreSQL in private subnet
- ElastiCache Redis in private subnet
- Security groups with least-privilege access
- CloudWatch log group
- IAM roles with minimal permissions
```

AI generates comprehensive Terraform with proper security configurations:

```
## VPC
module "vpc" {
  source  = "terraform-aws-modules/vpc/aws"
  version = "5.5.0"

  name = "${var.project_name}-vpc"
  cidr = "10.0.0.0/16"

  azs             = ["${var.aws_region}a", "${var.aws_region}b"]
  private_subnets = ["10.0.1.0/24", "10.0.2.0/24"]
  public_subnets  = ["10.0.101.0/24", "10.0.102.0/24"]
```

```
  enable_nat_gateway   = true
  single_nat_gateway   = var.environment != "production"
  enable_dns_hostnames = true

  tags = var.common_tags
}

## Security Group for ECS tasks
resource "aws_security_group" "ecs_tasks" {
  name_prefix = "${var.project_name}-ecs-"
  vpc_id      = module.vpc.vpc_id

  ingress {
    from_port       = var.container_port
    to_port         = var.container_port
    protocol        = "tcp"
    security_groups = [aws_security_group.alb.id]
  }

  egress {
    from_port   = 0
    to_port     = 0
    protocol    = "-1"
    cidr_blocks = ["0.0.0.0/0"]
  }
}

## RDS - private subnet only
resource "aws_db_instance" "main" {
  identifier     = "${var.project_name}-db"
  engine         = "postgres"
  engine_version = "16.1"
  instance_class = var.db_instance_class

  db_name  = var.db_name
  username = var.db_username
  password = var.db_password

  db_subnet_group_name   = aws_db_subnet_group.main.name
  vpc_security_group_ids = [aws_security_group.database.id]

  storage_encrypted = true
  multi_az          = var.environment == "production"
```

```
  backup_retention_period = var.environment == "production" ? 30 : 7
  skip_final_snapshot     = var.environment != "production"

  tags = var.common_tags
}
```

A Kubernetes Manifest Walkthrough

A concrete example of AI-assisted IaC, because the previous sections showed prompts and partial output. Here's what a full Kubernetes deployment looks like when you ask AI to generate it — and where you need to intervene.

Your prompt:

```
Generate Kubernetes manifests for a Node.js API service:
- Deployment: 3 replicas, rolling update (maxSurge 1, maxUnavailable 0)
- Resource requests: 256Mi memory, 250m CPU
- Resource limits: 512Mi memory, 500m CPU
- Liveness probe on /health, readiness probe on /ready
- Environment variables from a ConfigMap and a Secret
- Service: ClusterIP on port 80 targeting container port 3000
- HPA: scale 3-10 replicas based on 70% CPU utilization
```

AI generates:

```yaml
apiVersion: apps/v1
kind: Deployment
metadata:
  name: api-service
  labels:
    app: api-service
spec:
  replicas: 3
  strategy:
    type: RollingUpdate
    rollingUpdate:
      maxSurge: 1
      maxUnavailable: 0
  selector:
    matchLabels:
      app: api-service
  template:
    metadata:
      labels:
```

```yaml
        app: api-service
    spec:
      securityContext:
        runAsNonRoot: true
        runAsUser: 1001
        fsGroup: 1001
      containers:
        - name: api
          image: registry.example.com/api-service:latest
          ports:
            - containerPort: 3000
          resources:
            requests:
              memory: "256Mi"
              cpu: "250m"
            limits:
              memory: "512Mi"
              cpu: "500m"
          livenessProbe:
            httpGet:
              path: /health
              port: 3000
            initialDelaySeconds: 15
            periodSeconds: 20
            failureThreshold: 3
          readinessProbe:
            httpGet:
              path: /ready
              port: 3000
            initialDelaySeconds: 5
            periodSeconds: 10
          envFrom:
            - configMapRef:
                name: api-config
            - secretRef:
                name: api-secrets
---
apiVersion: v1
kind: Service
metadata:
  name: api-service
spec:
  type: ClusterIP
  selector:
    app: api-service
```

```yaml
  ports:
    - port: 80
      targetPort: 3000
---
apiVersion: autoscaling/v2
kind: HorizontalPodAutoscaler
metadata:
  name: api-service
spec:
  scaleTargetRef:
    apiVersion: apps/v1
    kind: Deployment
    name: api-service
  minReplicas: 3
  maxReplicas: 10
  metrics:
    - type: Resource
      resource:
        name: cpu
        target:
          type: Utilization
          averageUtilization: 70
```

This is solid. AI included the security context (non-root user), proper probe timing, and the correct HPA API version (`autoscaling/v2`, not the deprecated `v2beta2`). But here's what you should check and likely change:

1. **Image tag `latest`** is a problem. In production, always use a specific SHA or version tag. AI defaults to `latest` because most example code does.
2. **No PodDisruptionBudget.** With `maxUnavailable: 0` on the deployment strategy, you also need a PDB to prevent node drains from taking down all replicas simultaneously.
3. **No topology spread constraints.** If all three replicas land on the same node, a node failure takes the entire service down.

These are the production-readiness details AI includes when you ask for them but omits by default. The generated manifest is a strong starting point — it would take you 20 minutes to write from scratch, and AI gave it to you in seconds — but the review step is non-negotiable.

Monitoring and Observability

```
Create a monitoring setup with:
- Prometheus metrics for a Node.js application
- Grafana dashboard configuration (JSON model)
- Alert rules for: high error rate, high latency, low availability
- Structured logging configuration (Pino)
```

AI generates metrics instrumentation, dashboard definitions, and alert rules that would take hours to research and configure manually.

```
// Prometheus metrics for Node.js
import { register, Counter, Histogram, Gauge } from 'prom-client';

// Collect default Node.js metrics
import { collectDefaultMetrics } from 'prom-client';
collectDefaultMetrics();

// Custom application metrics
export const httpRequestDuration = new Histogram({
  name: 'http_request_duration_seconds',
  help: 'Duration of HTTP requests in seconds',
  labelNames: ['method', 'route', 'status_code'],
  buckets: [0.01, 0.05, 0.1, 0.5, 1, 2, 5],
});

export const httpRequestTotal = new Counter({
  name: 'http_requests_total',
  help: 'Total number of HTTP requests',
  labelNames: ['method', 'route', 'status_code'],
});

export const activeConnections = new Gauge({
  name: 'active_connections',
  help: 'Number of active connections',
});

export const dbQueryDuration = new Histogram({
  name: 'db_query_duration_seconds',
  help: 'Duration of database queries',
  labelNames: ['operation', 'table'],
  buckets: [0.001, 0.005, 0.01, 0.05, 0.1, 0.5, 1],
});
```

Terraform: A Quick IaC Workflow Tip

When using AI for Terraform, the highest-value workflow isn't generating resources from scratch — it's converting existing manually-created infrastructure into code. If you've been clicking through the AWS console, you have resources with no Terraform state. AI can help:

```
I created these AWS resources manually and need to bring them
under Terraform management:
- An S3 bucket named "myapp-uploads-prod"
- A CloudFront distribution pointing to that bucket
- An IAM role for Lambda with S3 read/write access

Generate the Terraform configuration that describes these existing
resources. Include terraform import commands for each one so I can
import the existing state without recreating anything.
```

AI generates both the HCL configuration and the import commands. This saves enormous time because Terraform's import workflow requires you to write the configuration first and then import — you can't import into an empty state. Having AI generate configuration to match existing resources lets you adopt IaC incrementally rather than via a big-bang migration.

Tip: After importing, run `terraform plan` and verify that Terraform shows no changes. If it shows drift, the AI-generated configuration doesn't match the actual resource. Adjust until the plan is clean.

What AI Gets Wrong in DevOps

Security defaults: AI generates functional but not always secure configurations. Always review: - Security group rules (is 0.0.0.0/0 necessary?) - IAM policies (least privilege?) - Encryption settings (at rest and in transit?) - Network exposure (should this be in a private subnet?)

Resource sizing: AI uses reasonable defaults, but your workload is unique. Monitor actual usage and adjust.

Cost implications: AI doesn't consider cost. A three-AZ, multi-NAT-gateway VPC with a large RDS instance might be correct for production but overkill for staging.

Vendor-specific quirks: Cloud providers have undocumented behaviors, region-specific limitations, and quota restrictions that AI may not know about.

Test infrastructure changes in a non-production environment first. AI-generated infrastructure that works in theory might have issues in practice.

Infrastructure deploys code. But when that code lives in a large, complex codebase with years of history and dozens of contributors, AI needs special strategies. Chapter 14 covers working with monorepos, legacy systems, and enterprise projects, where the sheer volume of code challenges AI's capabilities.

Part IV: Codebase and Infrastructure

"Any fool can write code that a computer can understand. Good programmers write code that humans can understand."

— Martin Fowler

Chapter 14: Working with Large Codebases

Strategies for AI in Monorepos, Legacy Code, Enterprise Projects

Most AI coding demos use small, clean example projects. Most real-world code lives in large, messy, interconnected codebases that have evolved over years. The gap between demo and reality is where developers get frustrated with AI — not because AI can't handle large codebases, but because the approach needs to change.

This chapter covers strategies for using AI effectively when your project is too large for any tool to grasp all at once.

The Context Window Problem

Every AI model has a context window — the maximum amount of text it can consider at once. Even the largest models in 2026 top out at around 1 million tokens, which sounds like a lot until you realize a medium-sized monorepo can easily contain 10 million tokens of source code.

AI can't read your entire codebase simultaneously. It reads portions, and the quality of its output depends on whether it has the right portions.

Strategy 1: Guide the AI to the Right Files

Instead of asking AI to figure out your codebase structure, tell it:

```
## Bad: AI has to guess where things are
claude "Fix the authentication bug"

## Good: AI knows exactly where to look
claude "There's a bug in the JWT validation in src/middleware/auth.ts.
  It's not checking token expiration. The token schema is defined
  in src/types/auth.ts and the auth configuration is in
  src/config/auth.config.ts."
```

Strategy 2: Use Project Maps

Create a high-level project map that AI can reference:

```
## PROJECT_MAP.md

### Architecture Overview
This is a monorepo managed by Turborepo containing:
- apps/web – Next.js frontend
- apps/api – Express backend API
- apps/admin – React admin dashboard
- apps/worker – Background job processor
- packages/shared – Shared TypeScript types and utilities
- packages/db – Prisma schema and database client
- packages/ui – Shared React component library
- packages/config – Shared ESLint, TypeScript, Tailwind configs

### Key Data Flows
1. User actions → web → api → db
2. Async operations → api → Redis queue → worker → db
3. Admin operations → admin → api → db

### Module Ownership
- Authentication: apps/api/src/auth/ + packages/shared/src/auth/
- Products: apps/api/src/modules/products/ + apps/web/src/features/products/
- Orders: apps/api/src/modules/orders/ + apps/worker/src/jobs/order-*.ts
- Payments: apps/api/src/modules/payments/ (Stripe integration)
```

Put this in your CLAUDE.md or a separate project map file. AI uses it to navigate your codebase without having to read every file.

Strategy 3: Scope Your Requests

For large codebases, narrow your requests to specific modules:

```
## Instead of this (too broad)
claude "Add error logging to the application"

## Do this (scoped to a module)
claude "Add structured error logging to the order processing module
  (apps/api/src/modules/orders/). Use our existing logger from
  packages/shared/src/logger.ts. Log order ID, user ID, error type,
  and stack trace for all caught exceptions."
```

Monorepo Strategies

Understanding Package Boundaries

AI needs to know which packages depend on which:

```
claude "Here's our monorepo dependency graph:
  - apps/web depends on packages/shared, packages/ui
  - apps/api depends on packages/shared, packages/db
  - apps/worker depends on packages/shared, packages/db
  - packages/ui depends on packages/shared
  - packages/db depends on packages/shared (for types)

  I need to add a new 'organization' concept. Where should the
  types go? Where should the database schema changes go?
  Where should the API endpoints go?"
```

Cross-Package Changes

When changes span multiple packages, break them into steps:

```
## Step 1: Schema and types
claude "Add an Organization model to the Prisma schema in
  packages/db/ and add the TypeScript types to packages/shared/src/types/"

## Step 2: API endpoints
claude "Add organization CRUD endpoints to apps/api/. Use the types
  from packages/shared and the Prisma client from packages/db."

## Step 3: Frontend
claude "Add organization management pages to apps/admin/.
  Use the types from packages/shared and UI components
  from packages/ui."
```

Each step stays within a manageable scope.

Legacy Code Strategies

Understanding Before Changing

The biggest mistake with AI and legacy code is jumping straight to changes. Build understanding first:

```
claude "Read the files in src/legacy/billing/ and explain:
1. What is the overall purpose of this module?
2. What are the entry points (functions called from outside)?
3. What external dependencies does it have?
4. What database tables does it access?
5. Are there any obvious bugs or code smells?
6. How is error handling done?

Do not suggest any changes yet — just help me understand it."
```

Incremental Modernization

For legacy code, the strangler fig pattern works well with AI:

```
## Phase 1: Create new interface alongside old
claude "Create a new BillingService class that exposes the same
  public interface as the legacy billing module. Internally,
  it should delegate to the legacy code. This is a facade."

## Phase 2: Migrate consumers one at a time
claude "Update the OrderController to use the new BillingService
  instead of importing directly from the legacy module."

## Phase 3: Replace internals gradually
claude "Now implement calculateInvoice in BillingService with
  clean code instead of delegating to the legacy function.
  Make sure it produces identical output for the same inputs."
```

Type Safety as a Bridge

Adding types to untyped legacy code is one of the highest-value AI tasks:

```
claude "Add TypeScript type definitions for the legacy billing
  module. Don't convert the JavaScript to TypeScript — just
  create a .d.ts declaration file that documents the actual
  types used. Infer types from the code's behavior."
```

This gives you type safety at the boundaries without rewriting the internals.

Enterprise Project Patterns

Working with Large Teams

When many developers use AI on the same codebase, coordination matters:

Standardize AI configuration. Every developer should use the same CLAUDE.md, .cursorrules, and copilot instructions. Commit these files and review them in PRs.

Establish AI code review norms. Decide as a team: does AI-generated code require different review standards? (Answer: same standards, but reviewers should be aware of common AI mistakes.)

Track AI-generated code. Some teams use commit message conventions (`[ai-assisted]`) to flag AI-generated code. This helps with debugging — if a bug is in AI-generated code, the failure mode might differ from a human bug.

Working with Internal Libraries

Enterprise codebases often have internal libraries that AI has never seen:

```
claude "We use an internal library called @acme/data-access for
  database operations. Here are the key interfaces and usage patterns:

  [paste relevant type definitions and a few usage examples]

  Now generate a new data access module for the Inventory entity
  following the same patterns."
```

Examples of correct usage beat documentation for internal tools.

Handling Proprietary Code

If your organization has concerns about sending proprietary code to AI services:

- **Claude Code**: Processes code locally and sends targeted queries to the API
- **Copilot Enterprise**: Runs on GitHub's infrastructure with enterprise data agreements
- **Tabnine**: Offers fully local model execution
- **Self-hosted solutions**: Some organizations run open-source models locally

Know your organization's data handling policies before using AI tools with proprietary code.

Performance Optimization for Large Projects

Targeted Analysis

```
claude "Our API response times have degraded. The p99 latency
  went from 200ms to 800ms over the last month. The most affected
  endpoints are:
  - GET /api/products (p99: 1200ms)
  - GET /api/orders/:id (p99: 900ms)

  Review the code for these endpoints and their downstream
  dependencies. Look for:
  - N+1 query patterns
  - Missing database indexes
  - Unnecessary data loading
  - Missing caching
  - Serialization overhead"
```

Dependency Analysis

```
claude "Analyze our package.json (and all workspace package.json files)
  and identify:
  1. Duplicate dependencies at different versions across workspaces
  2. Packages that are in dependencies but should be devDependencies
  3. Packages that are listed but never imported
  4. Packages with known security vulnerabilities
  5. Opportunities to replace multiple small packages with a single one"
```

Working with AI on Database-Heavy Codebases

Large applications often have complex database schemas with hundreds of tables:

```
claude "Here is our database schema (ERD or SQL):
  [paste or reference schema file]

  I need to add a reporting feature that shows:
  - Orders by product category, by month, for the last 12 months
  - Top 10 customers by revenue
  - Products with declining sales trends

  Design the queries. Consider:
  - These will run on a read replica
  - The orders table has 50M rows
```

```
- We need sub-second response times
- We can use materialized views if needed"
```

Practical Walkthrough: Onboarding to an Unfamiliar Codebase

Here's what it looks like to use AI to get productive in a codebase you've never seen before. This is one of the highest-value uses of AI for experienced developers — cutting the "where does anything live?" phase from days to hours.

Scenario: You just joined a team. The project is a Django application with 200+ files, and you need to fix a bug in the payment processing flow by end of week.

Step 1: Get the lay of the land.

```
claude "Give me a high-level overview of this project. What framework
  is it using? What's the directory structure? What are the main
  modules? Don't go into detail yet — just give me the map."
```

AI reads the project structure and reports back: Django 4.2, PostgreSQL, Celery for background tasks, Stripe for payments. It identifies the main Django apps: `accounts`, `products`, `orders`, `payments`, `notifications`. 30 seconds, not 15 minutes of clicking through directories.

Step 2: Understand the specific module you need.

```
claude "I need to understand the payment processing flow. Walk me
  through what happens when a user submits a payment, starting from
  the view that handles the form submission. Which files are involved?
  What's the sequence of calls?"
```

AI traces the flow: `payments/views.py` receives the POST, calls `payments/services.py` which creates a Stripe PaymentIntent, then a Celery task in `payments/tasks.py` polls for the result and updates the order status via `orders/services.py`. Now you know which files to read.

Step 3: Understand the data model.

```
claude "Show me the database models related to payments. How does
  a Payment relate to an Order? What statuses can a Payment have?
  Are there any model methods I should know about?"
```

AI reads the models and explains the relationships: `Payment` has a foreign key to `Order`, a `status` field with choices (pending, processing, succeeded, failed, refunded), and a `process()` method that handles the state machine transitions.

Step 4: Find the bug.

```
claude "The bug report says that some payments are being marked as
  'failed' even though Stripe shows them as succeeded. Look at the
  Celery task that polls Stripe for payment status. Is there a race
  condition or a status mapping issue?"
```

AI finds it: the Celery task checks the Stripe PaymentIntent status, but maps `requires_action` to `failed` instead of keeping it as `processing`. Some 3D Secure payments pass through `requires_action` temporarily before succeeding, and the task marks them as failed before the customer completes the 3D Secure challenge.

Total time from "I've never seen this code" to "I found the bug": about 30 minutes. Without AI, this would be a full day of reading code and tracing execution paths.

The key technique: Notice how each step built on the previous one. You didn't ask AI to "find the bug" immediately. You built context layer by layer — project overview, then module deep-dive, then data model, then targeted investigation. The layered approach works because it gives AI (and you) the right context at each step.

The Large Codebase Checklist

When working with AI in a large codebase:

1. **Have a CLAUDE.md or project map.** AI needs a guide to your codebase.
2. **Scope your requests.** Never ask AI to "fix the whole thing." Break it into modules.
3. **Provide context explicitly.** Tell AI which files are relevant.
4. **Use the right tool.** Claude Code and Cursor handle large codebases better than copy-paste chat tools.
5. **Verify dependencies.** After AI changes one module, check that dependent modules still work.
6. **Test in isolation first.** Run module-level tests before full suite tests.

7. **Document your AI interactions.** For complex changes, capture what you asked AI to do and why, so future developers understand the intent.

Working with AI in large codebases isn't harder — it just requires more intentional preparation. The payoff scales with codebase size: the bigger the project, the more AI saves you.

A Note on CLAUDE.md and Project Instructions

If you use Claude Code, the `CLAUDE.md` file at the root of your project is the single most impactful thing you can create for large-codebase productivity. It's read automatically at the start of every session, giving AI instant context about your project.

For large codebases, a good `CLAUDE.md` includes:

```
## Architecture
- Monorepo: apps/ for deployables, packages/ for shared code
- Backend: Express + Prisma, all routes in apps/api/src/routes/
- Frontend: Next.js, pages router (not app router)

## Conventions
- All database access through repository pattern (src/repositories/)
- Error handling: throw AppError subclasses, caught by global middleware
- Tests: colocated with source files as *.test.ts
- Imports: use @/ alias for absolute imports from src/

## Common Tasks
- Add a new API endpoint: create route in routes/, service in services/, reposit
ory in repositories/
- Add a new frontend page: create page in pages/, hook in hooks/, component in c
omponents/
- Database changes: edit prisma/schema.prisma, run npx prisma migrate dev

## Known Issues
- The legacy billing module (src/legacy/billing/) uses raw SQL - do not refactor
 without explicit request
- The auth middleware has two versions: v1 (session-based) and v2 (JWT) - new ro
utes use v2
```

This file costs 10 minutes to write and saves hours across every AI session in that project.

Chapter 15: Git Workflows with AI

Commits, PRs, Code Review Automation, Merge Conflict Resolution

Git is the universal constant of modern development. Every project uses it, every developer knows it (mostly), and every team has opinions about the "right" way to use it. AI smooths the rough edges — generating meaningful commit messages, creating well-structured pull requests, automating code review, and resolving merge conflicts without losing your mind.

AI-Generated Commit Messages

The Problem with Human Commit Messages

You know the pattern. It's 5 PM. You've been working on a feature all day. You stage your changes and type:

```
git commit -m "updates"
```

Or its cousins: "fix stuff," "WIP," "asdfgh," and the ever-popular "it works now."

AI eliminates this anti-pattern by analyzing your actual changes and generating meaningful commit messages.

Using Claude Code for Commits

```
## Claude Code analyzes your diff and generates a commit message
claude "Look at my staged changes and write a conventional commit
  message. Follow the format: type(scope): description.
  Include a body if the changes are complex."
```

Claude Code reads your diff and produces:

```
feat(auth): add JWT token refresh with sliding window expiration

- Add refresh token rotation on each use
- Implement sliding window expiration (extends on activity)
- Add token family tracking for reuse detection
```

```
- Update auth middleware to handle expired access tokens
  by attempting automatic refresh
```

Using Copilot for Commits

In VS Code with the Copilot extension, click the sparkle icon in the Source Control panel next to the commit message input. Copilot reads your staged changes and suggests a message.

Commit Message Best Practices with AI

Tell AI your commit conventions:

```
claude "Generate a commit message following these rules:
  - Conventional commits format (feat, fix, chore, docs, refactor, test)
  - Scope is the module name (auth, orders, products, etc.)
  - Subject line under 72 characters
  - Imperative mood ('add' not 'added')
  - Body explains WHY, not just WHAT
  - Reference issue numbers if visible in the code (TODO comments, etc.)
  - No emoji"
```

Pull Request Generation

Creating PRs with AI

AI can generate complete, well-structured pull request descriptions from your branch's changes:

```
## Using Claude Code
claude "Create a pull request. Compare my current branch against
  main and generate:
  - A clear title (under 72 characters)
  - A summary of what changed and why
  - A list of specific changes organized by category
  - Testing notes (what to test, how to test it)
  - Any deployment notes or breaking changes
  - Screenshots or UI descriptions if frontend changes exist"
```

Output:

```
### Add Bulk Product Import via CSV

#### Summary
```

```
Adds the ability for admin users to import products in bulk
via CSV upload. Validates the CSV format, shows a preview of
valid/invalid rows, and imports valid products in a background job.

#### Changes
**API**
- Add `POST /api/imports/upload` endpoint for CSV upload
- Add `GET /api/imports/:id` endpoint for import status polling
- Add `ImportJob` model and database migration

**Frontend**
- Add CSV upload component with drag-and-drop
- Add import preview table showing valid/invalid rows
- Add import progress indicator with real-time updates via SSE

**Background Jobs**
- Add `ProcessCSVImport` job with row-by-row validation
- Add retry logic for transient database errors

#### Testing
- Upload a valid CSV with 100 rows — all should import
- Upload a CSV with mixed valid/invalid rows — invalid rows
  should show errors, valid rows should import
- Upload a file that is not CSV — should show format error
- Upload a CSV exceeding 10MB — should be rejected

#### Deployment Notes
- Run database migration: `npm run db:migrate`
- No environment variable changes required
```

Using gh CLI with AI

```
## Generate PR description, then create PR
claude "Generate a PR description for my changes" > /tmp/pr-body.md
gh pr create --title "Add bulk product import" --body-file /tmp/pr-body.md
```

AI-Generated Commit Messages in Practice

Here's what this looks like with a real diff. Say you have staged changes across three files: you added a retry mechanism to the email service, updated the config to include retry settings, and added a test. The diff is 87 lines across `src/services/email.ts`, `src/config/email.config.ts`, and `tests/email.test.ts`.

AI sees the full diff and produces:

```
feat(email): add exponential backoff retry for transient send failures

- Retry up to 3 times on 429 and 5xx responses from the email provider
- Backoff schedule: 1s, 4s, 16s (exponential with jitter)
- Log each retry attempt with the error details and attempt number
- Non-retryable errors (4xx except 429) fail immediately
- Add EMAIL_MAX_RETRIES and EMAIL_BACKOFF_BASE to config

Closes #247
```

Compare that to what you'd have typed at 5 PM: "add email retries." Same diff, wildly different commit message. The AI version tells future developers (including future you) exactly what the retry strategy is, which errors trigger retries, and which issue it resolves.

AI-Generated PR Descriptions in Practice

PR descriptions benefit even more from AI because they need to communicate with reviewers who haven't been in your head all day. Here's a real example of the prompt and output.

After two days on a feature branch, you have eight commits adding a CSV export feature to an admin dashboard. You run:

```
claude "/pr"
```

Claude Code examines all eight commits, the full diff against main, and generates:

```
### Add CSV Export for Admin Reports

#### Summary
Admin users can now export any report table to CSV directly from the
dashboard. Exports run as background jobs to avoid blocking the UI
for large datasets (100K+ rows).

#### Changes
- **API:** New `POST /api/admin/exports` endpoint that queues a CSV
  generation job and returns a job ID
- **API:** New `GET /api/admin/exports/:id` endpoint that returns job
  status and download URL when complete
- **Worker:** New `GenerateCSVExport` BullMQ job that streams query
  results to a CSV file, uploads to S3, and generates a presigned URL
```

```
- **Frontend:** "Export CSV" button on all report pages with progress
  indicator and download link on completion
- **Database:** New `exports` table tracking export jobs (status,
  row count, file URL, expiration)

#### Testing
- Export a report with < 100 rows – should complete in < 2 seconds
- Export a report with 50K+ rows – should show progress and complete
  within 30 seconds
- Click export twice rapidly – should not create duplicate jobs
  (idempotency key based on report + filters + timestamp window)
- Download link should expire after 24 hours

#### Deployment Notes
- Run migration: `npm run db:migrate`
- New env vars: `EXPORT_S3_BUCKET`, `EXPORT_URL_EXPIRY_HOURS`
- BullMQ queue `csv-exports` must be registered in the worker config
```

This PR description would take 15–20 minutes to write manually. AI generates it in seconds because it has the complete diff. More importantly, it catches things you'd forget to mention — the new environment variables, the migration requirement, the BullMQ queue registration.

Automated Code Review

Setting Up AI Review on PRs

Several approaches work for automated AI review on pull requests:

GitHub Copilot for Pull Requests: Available with Copilot Enterprise. Reviews PRs automatically and leaves comments.

Claude Code in CI: Run Claude Code as a CI step to review PR changes:

```
## .github/workflows/ai-review.yml
name: AI Code Review

on:
  pull_request:
    types: [opened, synchronize]

jobs:
  review:
    runs-on: ubuntu-latest
```

```yaml
    permissions:
      pull-requests: write
    steps:
      - uses: actions/checkout@v4
        with:
          fetch-depth: 0

      - name: Get diff
        run: |
          git diff origin/main..HEAD > /tmp/diff.patch

      - name: AI Review
        run: |
          # Run your AI review tool of choice
          # Post results as PR comment
```

Third-Party Tools: CodeRabbit, Sourcery, and Ellipsis provide AI PR review as a service. Install them as GitHub Apps and they review every PR automatically.

What Automated Review Should Check

Configure your AI review to focus on high-value checks:

```
Review this PR diff for:
1. Security vulnerabilities (SQL injection, XSS, auth bypass)
2. Missing error handling
3. Potential null pointer dereferences
4. Performance issues (N+1 queries, missing indexes)
5. Violations of our coding standards (see CLAUDE.md)
6. Missing or inadequate tests for new functionality
7. Breaking changes to public APIs

Do NOT comment on:
- Style preferences that linters handle
- Import ordering
- Whitespace
```

Merge Conflict Resolution

Merge conflicts are tedious and error-prone. AI helps resolve them faster and more accurately.

Using AI to Resolve Conflicts

When you hit a conflict:

```
## See what conflicts exist
git diff --name-only --diff-filter=U

## Ask AI to resolve them
claude "I have merge conflicts in the following files:
  - src/services/order-service.ts
  - src/routes/orders.ts

  The merge is from feature/bulk-orders into main.
  The feature branch adds bulk order creation. Main had
  changes to order validation and pricing.

  Resolve the conflicts by keeping both sets of changes.
  The bulk order logic should use the updated validation
  and pricing from main."
```

Claude Code reads the conflicted files, understands both sides, and produces a resolution that preserves intent from both branches.

Complex Conflict Resolution

For conflicts in generated files (lock files, schemas):

```
## Prisma schema conflicts
claude "Resolve the merge conflict in prisma/schema.prisma.
  Both branches added new models — keep all models from both
  branches. If there are conflicting changes to existing models,
  the main branch version takes precedence for existing fields,
  but include new fields from the feature branch."
```

For lock files, it's almost always better to regenerate:

```
## Instead of resolving package-lock.json conflicts manually
git checkout --theirs package-lock.json
npm install
git add package-lock.json
```

Git History Analysis

AI can analyze your git history for insights:

```
## Find hotspots — files that change most often
git log --pretty=format: --name-only | sort | uniq -c | sort -rn | head -20

## Then ask AI to analyze the results
claude "These are the 20 most frequently changed files in our repo.
  Which ones might benefit from refactoring? High change frequency
  often indicates a file is doing too much or has unclear boundaries."
```

Understanding Old Changes

```
## What happened in this file over the last month?
git log --since="1 month ago" --oneline -- src/services/payment-service.ts

## Ask AI to summarize
claude "Here are the recent commits touching payment-service.ts:
  [paste git log output]

  Summarize the evolution of this file over the last month.
  Are there patterns that suggest problems (frequent bug fixes,
  feature churn, etc.)?"
```

Interactive Rebase with AI

When you need to clean up a branch's commit history before merging:

```
## Get the commit list
git log --oneline main..HEAD

## Ask AI for a rebase plan
claude "Here are my 12 commits on this branch:
  [paste git log output]

  Suggest how to squash and reorder these for a clean history.
  Group related commits together. Write the final commit messages."
```

AI might suggest:

```
Rebase plan:
1. Squash commits 1-3 into: "feat(auth): add JWT token refresh"
2. Squash commits 4-5 into: "feat(auth): add token family tracking"
3. Keep commit 6 as-is: "test(auth): add refresh token tests"
4. Squash commits 7-9 into: "fix(auth): handle expired token edge cases"
5. Drop commit 10 (temporary debug logging)
6. Squash commits 11-12 into: "docs(auth): update auth flow documentation"
```

Branch Management

```
## Find and clean up stale branches
claude "Run git branch -a and git log to identify branches that:
  1. Have been merged but not deleted
  2. Have not been updated in over 30 days
  3. Are behind main by more than 50 commits

  Suggest which to delete and which to rebase."
```

Git Hooks with AI

Set up pre-commit hooks that use AI:

```
#!/bin/bash
# .git/hooks/pre-commit

# Run AI review on staged changes
STAGED_DIFF=$(git diff --staged)
if [ -n "$STAGED_DIFF" ]; then
  echo "Running AI pre-commit check..."
  # Quick check for obvious issues
  echo "$STAGED_DIFF" | claude --quick "Check this diff for:
    - Hardcoded secrets (API keys, passwords, tokens)
    - console.log/print statements that should be removed
    - TODO comments without issue references
    Report only critical issues."
fi
```

What AI Gets Wrong with Git

Destructive suggestions: AI sometimes suggests force pushes, hard resets, and other destructive operations without warning. Understand what a git command does before running it

Rewriting shared history: AI might suggest rebasing commits that have been pushed to a shared branch. Never rewrite history that others have based work on.

Merge strategy confusion: AI might resolve conflicts by keeping only one side when both sides should be preserved. Review conflict resolutions carefully.

Commit scope: AI sometimes generates one large commit message for changes that should be split into multiple commits. Break large changes into logical units.

Overly verbose messages: AI can go too far the other way — generating commit messages four paragraphs long for a one-line fix. A good commit message is proportional to the change. A typo fix needs "fix: correct typo in error message," not a treatise on string accuracy. If AI generates a message longer than the diff it describes, ask it to shorten it.

Git is the most dangerous tool AI can automate because git errors can lose work. Run `git stash` or create a backup branch before following AI's git suggestions.

From Internal Versioning to Public Surface

Git tracks changes to code. APIs expose code to the world. Chapter 16 covers AI-assisted API development — designing, building, testing, and documenting APIs with AI tools that understand REST, GraphQL, and modern API patterns.

Chapter 16: API Development with AI

Designing, Building, Testing, and Documenting APIs

APIs are the contracts between systems. A well-designed API makes integration easy, evolution graceful, and debugging possible. A poorly designed API creates years of pain that gets worse with every consumer.

AI is exceptionally good at API development because APIs follow well-established patterns and conventions. REST, GraphQL, gRPC — domains where AI has seen millions of examples and produces output that follows best practices by default.

API Design with AI

REST API Design

Before writing any code, use AI to design your API surface:

```
Design a REST API for an e-commerce order management system.

Resources: Orders, OrderItems, Refunds, Shipments

For each resource, define:
- Endpoint paths (following REST conventions)
- HTTP methods
- Request/response schemas
- Status codes for success and error cases
- Pagination approach for list endpoints
- Filtering and sorting parameters

Constraints:
- Use consistent naming (camelCase for JSON fields)
- All list endpoints support cursor-based pagination
- All endpoints return the standard envelope: { data, error, meta }
- Use HTTP status codes correctly (don't return 200 for errors)
- Design for forward compatibility (additive changes should not break clients)
```

AI produces a comprehensive API design:

```
## Order Management API Design
```

```
### Orders

#### POST /api/v1/orders
Create a new order.
Request: { items: [{ productId, quantity }], shippingAddress: {...} }
Response (201): { data: Order, error: null, meta: {} }
Errors: 400 (validation), 409 (out of stock), 401 (unauthorized)

#### GET /api/v1/orders
List orders for the authenticated user.
Query: ?status=pending&sort=createdAt&order=desc&cursor=abc&limit=20
Response (200): { data: Order[], error: null, meta: { cursor, hasMore } }

#### GET /api/v1/orders/:orderId
Get a single order with items and shipment info.
Response (200): { data: Order, error: null, meta: {} }
Errors: 404 (not found), 403 (not your order)

#### PATCH /api/v1/orders/:orderId
Update order (only allowed in 'pending' status).
Request: { shippingAddress: {...} }
Response (200): { data: Order, error: null, meta: {} }
Errors: 404, 403, 409 (order not in pending status)

#### POST /api/v1/orders/:orderId/cancel
Cancel an order (action endpoint, not CRUD).
Request: { reason: "Changed my mind" }
Response (200): { data: Order, error: null, meta: {} }
Errors: 404, 403, 409 (order already shipped)
```

GraphQL Schema Design

```
Design a GraphQL schema for a project management application.

Core types: User, Project, Task, Comment, TimeEntry, Label

Requirements:
- Queries for listing and getting individual resources
- Mutations for CRUD operations on each resource
- Subscriptions for real-time task updates
- Connection-based pagination (Relay cursor specification)
- Input types for create/update operations with validation descriptions
- Proper nullable/non-nullable field design
- Custom scalars for DateTime and Duration
```

```
scalar DateTime
scalar Duration

type Query {
  me: User!
  project(id: ID!): Project
  projects(
    first: Int = 20
    after: String
    filter: ProjectFilter
  ): ProjectConnection!
  task(id: ID!): Task
}

type Mutation {
  createProject(input: CreateProjectInput!): CreateProjectPayload!
  updateProject(id: ID!, input: UpdateProjectInput!): UpdateProjectPayload!
  deleteProject(id: ID!): DeleteProjectPayload!

  createTask(input: CreateTaskInput!): CreateTaskPayload!
  updateTask(id: ID!, input: UpdateTaskInput!): UpdateTaskPayload!
  moveTask(id: ID!, status: TaskStatus!): MoveTaskPayload!
  assignTask(id: ID!, assigneeId: ID): AssignTaskPayload!
}

type Subscription {
  taskUpdated(projectId: ID!): Task!
  taskCreated(projectId: ID!): Task!
}

type Project {
  id: ID!
  name: String!
  description: String
  owner: User!
  members: [ProjectMember!]!
  tasks(
    first: Int = 20
    after: String
    filter: TaskFilter
  ): TaskConnection!
  createdAt: DateTime!
  updatedAt: DateTime!
}
```

```graphql
type Task {
  id: ID!
  title: String!
  description: String
  status: TaskStatus!
  priority: Priority!
  assignee: User
  labels: [Label!]!
  comments(first: Int = 10, after: String): CommentConnection!
  timeEntries: [TimeEntry!]!
  totalTimeSpent: Duration!
  dueDate: DateTime
  project: Project!
  createdBy: User!
  createdAt: DateTime!
  updatedAt: DateTime!
}

input CreateTaskInput {
  projectId: ID!
  title: String!
  description: String
  priority: Priority = MEDIUM
  assigneeId: ID
  labelIds: [ID!]
  dueDate: DateTime
}

type TaskConnection {
  edges: [TaskEdge!]!
  pageInfo: PageInfo!
  totalCount: Int!
}

enum TaskStatus {
  OPEN
  IN_PROGRESS
  IN_REVIEW
  DONE
  CANCELLED
}
```

Generating an OpenAPI Spec

One of the highest-leverage uses of AI in API development is generating OpenAPI (formerly Swagger) specifications. A well-written OpenAPI spec gives you documentation, client SDK generation, request validation, and mock servers — all from a single source of truth.

Here's the workflow. You have an existing API with a dozen endpoints but no spec. Instead of writing the spec by hand (which nobody enjoys), point AI at your route definitions:

```
Here are our Express route definitions for the Orders API:
[paste your routes file]

Generate a complete OpenAPI 3.1 specification. Include:
- Accurate request/response schemas matching our Zod validators
- Proper error responses for each endpoint (400, 401, 403, 404, 409)
- Authentication scheme (Bearer JWT)
- Pagination parameters for list endpoints
- Example values for all fields
```

AI generates:

```yaml
openapi: 3.1.0
info:
  title: Orders API
  version: 1.0.0
  description: Order management endpoints for the e-commerce platform

servers:
  - url: https://api.example.com/v1
    description: Production
  - url: http://localhost:3000/v1
    description: Development

security:
  - bearerAuth: []

paths:
  /orders:
    get:
      summary: List orders
      operationId: listOrders
      tags: [Orders]
```

```yaml
  parameters:
    - name: status
      in: query
      schema:
        type: string
        enum: [pending, processing, shipped, delivered, cancelled]
      example: pending
    - name: cursor
      in: query
      schema:
        type: string
      description: Pagination cursor from previous response
    - name: limit
      in: query
      schema:
        type: integer
        minimum: 1
        maximum: 100
        default: 20
  responses:
    '200':
      description: Paginated list of orders
      content:
        application/json:
          schema:
            type: object
            properties:
              data:
                type: array
                items:
                  $ref: '#/components/schemas/Order'
              error:
                type: 'null'
              meta:
                $ref: '#/components/schemas/PaginationMeta'
    '401':
      $ref: '#/components/responses/Unauthorized'

post:
  summary: Create a new order
  operationId: createOrder
  tags: [Orders]
  requestBody:
    required: true
    content:
```

```yaml
          application/json:
            schema:
              $ref: '#/components/schemas/CreateOrderInput'
            example:
              items:
                - productId: "prod_abc123"
                  quantity: 2
              shippingAddress:
                line1: "123 Main St"
                city: "Portland"
                state: "OR"
                zip: "97201"
      responses:
        '201':
          description: Order created successfully
          content:
            application/json:
              schema:
                type: object
                properties:
                  data:
                    $ref: '#/components/schemas/Order'
                  error:
                    type: 'null'
                  meta:
                    type: object
        '400':
          $ref: '#/components/responses/ValidationError'
        '409':
          description: Insufficient stock
          content:
            application/json:
              schema:
                $ref: '#/components/schemas/ErrorResponse'

components:
  securitySchemes:
    bearerAuth:
      type: http
      scheme: bearer
      bearerFormat: JWT

  schemas:
    Order:
      type: object
```

```yaml
      required: [id, status, items, total, createdAt]
      properties:
        id:
          type: string
          format: uuid
          example: "ord_a1b2c3d4"
        status:
          type: string
          enum: [pending, processing, shipped, delivered, cancelled]
        items:
          type: array
          items:
            $ref: '#/components/schemas/OrderItem'
        total:
          type: string
          pattern: '^\d+\.\d{2}$'
          example: "149.98"
        createdAt:
          type: string
          format: date-time
```

The generated spec is immediately usable. Paste it into Swagger Editor to visualize it, run `openapi-generator` to create client SDKs, or plug it into Express middleware like `express-openapi-validator` to validate every request and response against the spec automatically.

What AI gets wrong with OpenAPI: it sometimes generates schemas that are too loose — using `type: object` with `additionalProperties: true` when your actual API returns a fixed set of fields. It also tends to forget `required` arrays on object schemas, making every field appear optional. Review the schemas against your actual Zod validators or TypeScript types.

API Implementation Patterns

Versioning

```
claude "Add API versioning to our Express application.
  Requirements:
  - URL-based versioning (/api/v1/, /api/v2/)
  - V1 and V2 can coexist
  - V2 endpoints fall back to V1 if not explicitly defined
  - Version extraction middleware
  - Deprecation headers for V1 endpoints"
```

Error Handling

AI generates consistent error handling that's often better than what developers write manually:

```
// AI-generated error hierarchy for APIs
export class ApiError extends Error {
  constructor(
    public statusCode: number,
    public code: string,
    message: string,
    public details?: Record<string, unknown>
  ) {
    super(message);
    this.name = 'ApiError';
  }
}

export class ValidationError extends ApiError {
  constructor(errors: { field: string; message: string }[]) {
    super(400, 'VALIDATION_ERROR', 'Request validation failed', { errors });
  }
}

export class NotFoundError extends ApiError {
  constructor(resource: string, id: string) {
    super(404, 'NOT_FOUND', `${resource} with id '${id}' not found`);
  }
}

export class ConflictError extends ApiError {
  constructor(message: string) {
    super(409, 'CONFLICT', message);
  }
}

export class RateLimitError extends ApiError {
  constructor(retryAfterSeconds: number) {
    super(429, 'RATE_LIMITED', 'Too many requests', { retryAfterSeconds });
  }
}
```

Rate Limiting

```
claude "Implement API rate limiting with:
  - Per-API-key limits (different tiers: free=100/hr, pro=1000/hr, enterprise=10
000/hr)
  - Per-endpoint limits (expensive endpoints get lower limits)
  - Redis-backed sliding window counter
  - X-RateLimit-Limit, X-RateLimit-Remaining, X-RateLimit-Reset headers
  - 429 response with Retry-After header when limited
  - Bypass for internal service-to-service calls"
```

API Testing

Contract Testing

```
claude "Generate contract tests for our Orders API using Pact.
  The consumer is the frontend React app. The provider is the
  Express API. Test these interactions:
  - Consumer creates an order and expects a 201 with order data
  - Consumer lists orders and expects paginated results
  - Consumer cancels an order and expects updated status
  - Consumer requests a non-existent order and expects 404"
```

Load Testing

```
claude "Generate a k6 load test script for our API:
  - Scenario 1: Normal load (50 VUs, 5 minutes)
  - Scenario 2: Spike test (ramp to 500 VUs over 1 minute)
  - Scenario 3: Soak test (100 VUs, 1 hour)

  Test endpoints:
  - GET /api/v1/products (list, with pagination)
  - POST /api/v1/orders (create order, requires auth)
  - GET /api/v1/orders/:id (get order details)

  Include proper authentication (get a token first).
  Set thresholds: p95 < 500ms, error rate < 1%."
```

```
import http from 'k6/http';
import { check, sleep } from 'k6';
import { Rate } from 'k6/metrics';

const errorRate = new Rate('errors');
```

```
const BASE_URL = __ENV.API_URL || 'http://localhost:3000';

export const options = {
  scenarios: {
    normal_load: {
      executor: 'constant-vus',
      vus: 50,
      duration: '5m',
    },
    spike_test: {
      executor: 'ramping-vus',
      startVUs: 0,
      stages: [
        { duration: '1m', target: 500 },
        { duration: '2m', target: 500 },
        { duration: '1m', target: 0 },
      ],
      startTime: '6m',
    },
  },
  thresholds: {
    http_req_duration: ['p(95)<500'],
    errors: ['rate<0.01'],
  },
};

function getAuthToken() {
  const res = http.post(`${BASE_URL}/api/v1/auth/login`, JSON.stringify({
    email: 'loadtest@example.com',
    password: 'loadtest123',
  }), { headers: { 'Content-Type': 'application/json' } });

  return JSON.parse(res.body).data.token;
}

export function setup() {
  return { token: getAuthToken() };
}

export default function (data) {
  const headers = {
    'Content-Type': 'application/json',
    'Authorization': `Bearer ${data.token}`,
  };
```

```
  // List products
  const listRes = http.get(`${BASE_URL}/api/v1/products?limit=20`, { headers });
  check(listRes, { 'list products 200': (r) => r.status === 200 });
  errorRate.add(listRes.status !== 200);

  sleep(1);

  // Create order
  const orderRes = http.post(`${BASE_URL}/api/v1/orders`, JSON.stringify({
    items: [{ productId: 'test-product-1', quantity: 1 }],
  }), { headers });
  check(orderRes, { 'create order 201': (r) => r.status === 201 });
  errorRate.add(orderRes.status !== 201);

  if (orderRes.status === 201) {
    const orderId = JSON.parse(orderRes.body).data.id;

    // Get order
    const getRes = http.get(`${BASE_URL}/api/v1/orders/${orderId}`, { headers });
    check(getRes, { 'get order 200': (r) => r.status === 200 });
    errorRate.add(getRes.status !== 200);
  }

  sleep(1);
}
```

API Documentation Generation

Covered in depth in Chapter 9, but specific to APIs:

```
claude "Generate an interactive API documentation page from our
  OpenAPI spec. Include:
  - Endpoint listing grouped by resource
  - Try-it-out functionality with curl commands
  - Authentication guide
  - Rate limiting information
  - Error code reference
  - SDK code examples in Python, JavaScript, and Go"
```

SDK Generation

For APIs that need client libraries:

```
claude "Generate a TypeScript SDK client for our Orders API.
  Requirements:
  - Type-safe methods for each endpoint
  - Automatic authentication header injection
  - Built-in retry with exponential backoff
  - Request/response interceptors
  - Pagination helpers (auto-follow cursors)
  - Error types that match our API error codes
  - Zero runtime dependencies (use native fetch)"

export class OrdersClient {
  constructor(private config: ClientConfig) {}

  async listOrders(params?: ListOrdersParams): Promise<PaginatedResponse<Order>>
 {
    return this.request<PaginatedResponse<Order>>('GET', '/api/v1/orders', {
      query: params,
    });
  }

  async getOrder(orderId: string): Promise<Order> {
    const response = await this.request<ApiResponse<Order>>(
      'GET', `/api/v1/orders/${orderId}`
    );
    return response.data;
  }

  async createOrder(input: CreateOrderInput): Promise<Order> {
    const response = await this.request<ApiResponse<Order>>(
      'POST', '/api/v1/orders', { body: input }
    );
    return response.data;
  }

  // Pagination helper
  async *listAllOrders(params?: Omit<ListOrdersParams, 'cursor'>): AsyncGenerato
r<Order> {
    let cursor: string | undefined;
    do {
      const response = await this.listOrders({ ...params, cursor });
      for (const order of response.data) {
        yield order;
      }
      cursor = response.meta.cursor;
    } while (cursor);
```

```
  }

  private async request<T>(
    method: string,
    path: string,
    options?: RequestOptions
  ): Promise<T> {
    // Implementation with retry, auth, and error handling
  }
}
```

What AI Gets Wrong with APIs

Inconsistent naming: AI might mix camelCase and snake_case within the same API. Specify your convention explicitly.

Over-complex responses: AI sometimes nests data too deeply or includes unnecessary metadata. Flatten where possible.

Missing idempotency: AI rarely adds idempotency keys to mutation endpoints unless you ask. For any endpoint that modifies state, consider requiring an idempotency key.

Wrong HTTP methods: AI sometimes uses POST for operations that should be PUT or PATCH, or uses DELETE for operations that should be a POST to a cancellation endpoint. A good rule: if the operation isn't CRUD, it probably needs an action endpoint (`POST /orders/:id/cancel`) rather than mapping awkwardly onto HTTP methods.

No versioning strategy: AI generates `/api/orders` when you should be generating `/api/v1/orders` from the start. Retrofitting API versioning is painful. Include it in your initial prompt.

Ignoring backwards compatibility: AI might suggest changes to existing endpoints that would break clients. Always ask: "Would this break existing consumers?"

API design has lasting consequences. Review AI-generated API designs carefully before implementing them. Changing an API after consumers depend on it is expensive.

APIs exchange data with the outside world, which makes security essential. Chapter 17 covers security-aware AI development — the specific security risks AI-generated code introduces and how to prevent them. If you use AI to generate code that handles user data, authentication, or network communication, treat that chapter as required reading.

Part V: Specialized Development

"The most disastrous thing that you can ever learn is your first programming language."

— Alan Kay

Chapter 17: Security-Aware AI Development

Avoiding AI-Introduced Vulnerabilities, Security Scanning

This is the chapter I need you to take seriously.

AI-generated code has security vulnerabilities. Not occasionally. Routinely. Studies consistently show that developers using AI code generation produce code with more security vulnerabilities than developers working without AI, primarily because AI makes it easy to generate large volumes of code that looks correct but has subtle security flaws.

The solution isn't to stop using AI. The solution is to understand how AI-generated code fails from a security perspective and to build verification into your workflow.

The AI Security Problem

AI coding tools are trained on vast repositories of public code. That code includes countless examples of insecure patterns — SQL injection, hardcoded credentials, broken authentication, missing input validation. AI doesn't distinguish between secure and insecure code. It learns both patterns and reproduces whichever one matches the context of your request.

Here's what makes this dangerous: AI-generated insecure code often looks correct to a human reviewer. It compiles, it passes basic tests, and it does what you asked. The vulnerability is in what it doesn't do — the missing validation, the absent authorization check, the SQL query formatted instead of parameterized.

The Top 10 AI-Generated Vulnerabilities

1. SQL Injection

The most common and most dangerous. AI regularly generates string-interpolated SQL:

```python
## AI generates this constantly (VULNERABLE)
@app.route('/users/<username>')
def get_user(username):
    query = f"SELECT * FROM users WHERE username = '{username}'"
    result = db.execute(query)
    return jsonify(result)
## Input: ' OR '1'='1  →  Returns ALL users
## Input: '; DROP TABLE users; --  →  Deletes the table

## SECURE version
@app.route('/users/<username>')
def get_user(username):
    result = db.execute(
        text("SELECT * FROM users WHERE username = :name"),
        {"name": username}
    )
    return jsonify(result)
```

Always check: Is every database query parameterized? Are there any string concatenations or f-strings in SQL?

2. Cross-Site Scripting (XSS)

AI-generated frontend code often injects user content without escaping:

```javascript
// VULNERABLE — AI generates this for "display user content"
function Comment({ comment }) {
  return <div dangerouslySetInnerHTML={{ __html: comment.body }} />;
  // If comment.body contains <script>alert('xss')</script>, it executes
}

// SECURE
function Comment({ comment }) {
  return <div>{comment.body}</div>;  // React auto-escapes by default
}

// If you NEED HTML rendering (rich text editor content):
import DOMPurify from 'dompurify';
function Comment({ comment }) {
  return <div dangerouslySetInnerHTML={{
    __html: DOMPurify.sanitize(comment.body)
  }} />;
}
```

3. Missing Authentication

AI generates functional endpoints and forgets authentication:

```
// AI generates this (NO AUTH CHECK)
router.delete('/api/users/:id', async (req, res) => {
  await prisma.user.delete({ where: { id: req.params.id } });
  res.status(204).end();
});

// SECURE — with auth AND authorization
router.delete('/api/users/:id',
  requireAuth,           // Is the user logged in?
  requireRole('admin'),  // Is the user an admin?
  async (req, res) => {
    await prisma.user.update({
      where: { id: req.params.id },
      data: { deletedAt: new Date() },  // Soft delete, not hard delete
    });
    res.status(204).end();
  }
);
```

4. Missing Authorization (IDOR)

Even when authentication exists, AI often skips authorization — checking that the authenticated user has permission to access the specific resource:

```
## VULNERABLE — authenticated but no authorization check
@app.route('/api/orders/<order_id>')
@require_auth
def get_order(order_id):
    order = db.orders.find_one(order_id)
    return jsonify(order)
    # Any authenticated user can view ANY order by guessing IDs

## SECURE
@app.route('/api/orders/<order_id>')
@require_auth
def get_order(order_id):
    order = db.orders.find_one(order_id)
    if order is None:
        abort(404)
    if order.user_id != current_user.id and not current_user.is_admin:
```

```
        abort(403)
    return jsonify(order)
```

This is called Insecure Direct Object Reference (IDOR), and it's one of the most common vulnerabilities in web applications. AI almost never adds these checks unless explicitly asked.

5. Hardcoded Secrets

AI training data is full of example code with hardcoded API keys:

```
// AI generates this (DANGEROUS)
const stripe = new Stripe('sk_live_abc123xyz789');

const jwtSecret = 'super-secret-key-change-me';

const dbConfig = {
  host: 'prod-db.example.com',
  password: 'admin123',
};

// SECURE — environment variables
const stripe = new Stripe(process.env.STRIPE_SECRET_KEY!);
const jwtSecret = process.env.JWT_SECRET!;
const dbConfig = {
  host: process.env.DB_HOST!,
  password: process.env.DB_PASSWORD!,
};
```

6. Insecure Defaults

AI generates code that works with the most permissive configuration:

```
// VULNERABLE — overly permissive CORS
app.use(cors({
  origin: '*',                    // Any origin can access
  credentials: true,              // Sends cookies cross-origin
}));

// SECURE
app.use(cors({
  origin: process.env.ALLOWED_ORIGINS?.split(',') || [],
  credentials: true,
  methods: ['GET', 'POST', 'PUT', 'DELETE'],
```

```
  allowedHeaders: ['Content-Type', 'Authorization'],
}));
```

7. Missing Input Validation

AI generates endpoints that trust user input:

```
// VULNERABLE — no validation
app.post('/api/users', async (req, res) => {
  const user = await db.users.create(req.body);  // Whatever the user sends
  res.json(user);
});

// SECURE — validate everything
const createUserSchema = z.object({
  email: z.string().email().max(255),
  name: z.string().min(1).max(100),
  role: z.enum(['user', 'moderator']),  // Not 'admin'!
});

app.post('/api/users', async (req, res) => {
  const validated = createUserSchema.parse(req.body);
  const user = await db.users.create(validated);
  res.json(user);
});
```

8. Path Traversal

When handling file operations:

```
## VULNERABLE
@app.route('/files/<filename>')
def download_file(filename):
    return send_file(f'/uploads/{filename}')
    # Input: ../../../etc/passwd → System file exposure

## SECURE
import os
from werkzeug.utils import secure_filename

@app.route('/files/<filename>')
def download_file(filename):
    safe_name = secure_filename(filename)
    filepath = os.path.join('/uploads', safe_name)
    if not os.path.abspath(filepath).startswith('/uploads/'):
```

```
        abort(400)
    return send_file(filepath)
```

9. Information Leakage

AI returns verbose error messages that expose internals:

```
// VULNERABLE — leaks stack trace and database info
app.use((err, req, res, next) => {
  res.status(500).json({
    error: err.message,
    stack: err.stack,
    query: err.query,  // Database query visible to user!
  });
});

// SECURE
app.use((err, req, res, next) => {
  logger.error({ err, requestId: req.id });  // Log details internally

  if (err instanceof AppError) {
    res.status(err.statusCode).json({
      error: { code: err.code, message: err.message },
    });
  } else {
    res.status(500).json({
      error: { code: 'INTERNAL_ERROR', message: 'An unexpected error occurred' },
    });
  }
});
```

10. Cryptographic Mistakes

AI generates weak cryptographic code:

```
## VULNERABLE — weak hashing
import hashlib
password_hash = hashlib.md5(password.encode()).hexdigest()  # MD5 is broken

## Also VULNERABLE — no salt
password_hash = hashlib.sha256(password.encode()).hexdigest()

## SECURE — use a proper password hashing library
import bcrypt
password_hash = bcrypt.hashpw(password.encode(), bcrypt.gensalt(rounds=12))
```

Real AI-Introduced Vulnerabilities: A Case Study

Here's a concrete example of how AI introduces a vulnerability chain — not just one isolated mistake, but a sequence of plausible-looking code that compounds into a serious security issue.

You ask AI to build a password reset flow. Here's what a typical AI generates:

```
// AI-generated password reset (MULTIPLE VULNERABILITIES)
router.post('/api/auth/forgot-password', async (req, res) => {
  const { email } = req.body;
  const user = await prisma.user.findUnique({ where: { email } });

  if (!user) {
    return res.status(404).json({ error: 'User not found' });  // BUG 1
  }

  const resetToken = Math.random().toString(36).substring(7);   // BUG 2

  await prisma.passwordReset.create({
    data: {
      userId: user.id,
      token: resetToken,                                          // BUG 3
      expiresAt: new Date(Date.now() + 24 * 60 * 60 * 1000),    // BUG 4
    },
  });

  await sendEmail(user.email, `Reset your password:
    https://example.com/reset?token=${resetToken}`);

  res.json({ message: `Reset link sent to ${email}` });          // BUG 5
});
```

Five security bugs in one endpoint, and every line looks reasonable to a developer who is not thinking about security.

Bug 1: User enumeration. Returning 404 for unknown emails tells attackers which emails are registered. The response should be identical regardless of whether the email exists.

Bug 2: Weak token generation. `Math.random()` isn't cryptographically secure. An attacker who knows when the request was made can predict the token. Use `crypto.randomBytes(32).toString('hex')` instead.

Bug 3: Token stored in plaintext. If the database is compromised, the attacker has every active reset token. Store a hash of the token (just like passwords).

Bug 4: 24-hour expiration is too long. Password reset tokens should expire in 15–60 minutes. A 24-hour window gives attackers an entire day to intercept or brute-force the token.

Bug 5: Email echo. Returning the email in the response confirms to an attacker that the email exists, even if you fix Bug 1.

Here's the secure version:

```
import { randomBytes, createHash } from 'crypto';

router.post('/api/auth/forgot-password', async (req, res) => {
  const { email } = req.body;
  const user = await prisma.user.findUnique({ where: { email } });

  if (user) {
    const resetToken = randomBytes(32).toString('hex');
    const tokenHash = createHash('sha256').update(resetToken).digest('hex');

    // Invalidate any existing reset tokens for this user
    await prisma.passwordReset.deleteMany({ where: { userId: user.id } });

    await prisma.passwordReset.create({
      data: {
        userId: user.id,
        tokenHash,
        expiresAt: new Date(Date.now() + 30 * 60 * 1000), // 30 minutes
      },
    });

    await sendEmail(user.email,
      `Reset your password: https://example.com/reset?token=${resetToken}`);
  }

  // Same response regardless of whether user exists
  res.json({ message: 'If an account exists, a reset link has been sent.' });
});
```

The point isn't that AI is bad at security. The point is that AI optimizes for functionality, and security is the set of things a functional implementation doesn't do. AI built a working password reset. It just didn't build a safe one.

The Security Review Checklist

Run this checklist on every AI-generated code change:

```
□ Authentication: Is authentication required on all non-public endpoints?
□ Authorization: Does each endpoint check that the user can access
  this specific resource?
□ Input validation: Is all user input validated before use?
□ SQL injection: Are all database queries parameterized?
□ XSS: Is user-generated content escaped before rendering?
□ CSRF: Are state-changing operations protected against CSRF?
□ Secrets: Are all secrets in environment variables, not code?
□ CORS: Is CORS configured with specific origins, not wildcards?
□ Error handling: Do error responses avoid leaking internal details?
□ File handling: Are file paths sanitized against path traversal?
□ Cryptography: Are passwords hashed with bcrypt/argon2, not MD5/SHA?
□ Rate limiting: Are expensive or sensitive endpoints rate limited?
□ Logging: Are sensitive values (passwords, tokens) excluded from logs?
□ HTTPS: Is HTTPS enforced? Are secure cookie flags set?
□ Dependencies: Are third-party packages from legitimate sources?
```

Automated Security Scanning

Static Analysis

Run security scanners on AI-generated code:

```
## Semgrep — finds security antipatterns
semgrep --config auto src/

## npm audit — checks dependencies
npm audit

## Bandit — Python security linter
bandit -r src/

## gosec — Go security checker
gosec ./...
```

Integration with AI Review

```
claude "Run a security review on the changes in my current branch.
  Check for OWASP Top 10 vulnerabilities:
  1. Injection (SQL, NoSQL, command injection)
  2. Broken authentication
  3. Sensitive data exposure
  4. XML external entities (XXE)
  5. Broken access control
  6. Security misconfiguration
  7. Cross-site scripting (XSS)
  8. Insecure deserialization
  9. Using components with known vulnerabilities
  10. Insufficient logging and monitoring

  Be specific about file names and line numbers."
```

Security-First Prompting

When generating code that handles sensitive operations, be explicit about security requirements:

```
Create a user registration endpoint with these SECURITY requirements:
- Hash passwords with bcrypt (cost factor 12)
- Validate email format AND check for disposable email providers
- Rate limit: 5 registration attempts per IP per hour
- Send verification email (don't activate account until verified)
- Don't reveal whether an email is already registered (timing-safe comparison)
- Log registration attempts (without logging the password)
- Return consistent response time regardless of outcome (prevent timing attacks)
- Set secure headers (HSTS, CSP, X-Frame-Options)
```

When you specify security requirements explicitly, AI generates much more secure code. The problem comes when you don't mention security and AI defaults to the fastest, most functional (but insecure) pattern.

The Dependency Trap

One vulnerability category deserves its own callout because it's unique to AI-generated code: dependency risk. When AI suggests a library to solve a problem, it doesn't check whether that library is actively maintained, has known vulnerabilities, or even still exists.

```
## AI suggests a package that was popular in 2022 but abandoned in 2024
npm install cool-auth-helper
## 6 known CVEs, last published 2 years ago, no maintainer
```

Run `npm audit` (or equivalent) after installing AI-suggested packages. Check the package's GitHub — when was the last commit? Are issues being addressed? How many maintainers does it have? A single-maintainer package with no activity for 12 months is a supply chain risk.

Better yet, tell AI your constraints:

```
I need to validate JWTs. Suggest a library that:
- Has been updated within the last 6 months
- Has at least 1000 GitHub stars
- Is maintained by an organization, not an individual
- Has zero known CVEs
Or tell me if I should just use the built-in jsonwebtoken package.
```

AI will still sometimes suggest outdated packages, but explicit constraints reduce the frequency.

Security Training Your AI Tools

Add security requirements to your project configuration:

```
## In CLAUDE.md or .cursorrules

### Security Requirements
- All database queries MUST use parameterized queries
- All endpoints MUST have authentication middleware unless explicitly public
- All endpoints MUST include authorization checks (not just authentication)
- All user input MUST be validated using Zod schemas
- Passwords MUST be hashed with bcrypt (cost 12+)
- API keys and secrets MUST come from environment variables
- Error responses MUST NOT include stack traces or internal details
- CORS MUST specify explicit allowed origins
- All file paths MUST be sanitized against path traversal
- Logging MUST exclude passwords, tokens, and PII
```

With these rules in your project context, AI tools follow them by default instead of generating insecure patterns.

Chapter 18: AI-Assisted Database Work

Schema Design, Query Optimization, Migrations

Databases underpin nearly every application, and database mistakes are among the most expensive to fix. A bad schema or a missing index in development becomes a production outage at scale. A poorly planned migration causes hours of downtime.

AI is a strong partner for database work. It knows SQL deeply, understands query optimization, generates correct migration files, and reads EXPLAIN output. This chapter covers AI-assisted database work end to end — from initial design to production optimization.

Schema Design

Starting from Requirements

```
Design a PostgreSQL schema for a multi-tenant SaaS helpdesk application.

Core entities:
- Tenants (organizations using the platform)
- Users (belong to one tenant, have roles)
- Tickets (created by users, assigned to agents)
- Comments (on tickets, by users or agents)
- Tags (per-tenant, applied to tickets, many-to-many)
- Attachments (on tickets or comments)
- SLA policies (per-tenant, define response/resolution times)

Requirements:
- Row-level security for tenant isolation
- Soft deletes on all entities
- Audit columns (created_at, updated_at, created_by, updated_by)
- Full-text search on tickets (title + description)
- Ticket status workflow: open → in_progress → waiting → resolved → closed
- Indexes for common queries (list by status, search, recent activity)
- Consider partitioning for tickets table (large dataset)
```

```
Output as SQL DDL with comments explaining design decisions.
```

AI generates comprehensive schemas with thoughtful design:

```
-- Tenants
CREATE TABLE tenants (
    id UUID PRIMARY KEY DEFAULT gen_random_uuid(),
    name TEXT NOT NULL,
    slug TEXT NOT NULL UNIQUE,
    plan TEXT NOT NULL DEFAULT 'free' CHECK (plan IN ('free', 'pro', 'enterprise
')),
    settings JSONB NOT NULL DEFAULT '{}',
    created_at TIMESTAMPTZ NOT NULL DEFAULT now(),
    deleted_at TIMESTAMPTZ
);

-- Row-Level Security: every table includes tenant_id
-- and RLS policies enforce isolation
ALTER TABLE tenants ENABLE ROW LEVEL SECURITY;

-- Tickets with full-text search
CREATE TABLE tickets (
    id UUID PRIMARY KEY DEFAULT gen_random_uuid(),
    tenant_id UUID NOT NULL REFERENCES tenants(id),
    ticket_number SERIAL,  -- Human-readable, per-tenant
    title TEXT NOT NULL,
    description TEXT,
    status TEXT NOT NULL DEFAULT 'open'
        CHECK (status IN ('open', 'in_progress', 'waiting', 'resolved', 'closed
')),
    priority TEXT NOT NULL DEFAULT 'medium'
        CHECK (priority IN ('low', 'medium', 'high', 'urgent')),
    requester_id UUID NOT NULL REFERENCES users(id),
    assignee_id UUID REFERENCES users(id),
    first_response_at TIMESTAMPTZ,  -- For SLA tracking
    resolved_at TIMESTAMPTZ,
    created_at TIMESTAMPTZ NOT NULL DEFAULT now(),
    updated_at TIMESTAMPTZ NOT NULL DEFAULT now(),
    created_by UUID NOT NULL REFERENCES users(id),
    deleted_at TIMESTAMPTZ,

    -- Full-text search vector
    search_vector TSVECTOR GENERATED ALWAYS AS (
        setweight(to_tsvector('english', coalesce(title, '')), 'A') ||
```

```
        setweight(to_tsvector('english', coalesce(description, '')), 'B')
    ) STORED,

    -- Unique ticket number per tenant
    UNIQUE (tenant_id, ticket_number)
) PARTITION BY RANGE (created_at);

-- Partition by quarter for large datasets
CREATE TABLE tickets_2026_q1 PARTITION OF tickets
    FOR VALUES FROM ('2026-01-01') TO ('2026-04-01');
CREATE TABLE tickets_2026_q2 PARTITION OF tickets
    FOR VALUES FROM ('2026-04-01') TO ('2026-07-01');

-- Indexes optimized for common queries
CREATE INDEX idx_tickets_tenant_status ON tickets(tenant_id, status)
    WHERE deleted_at IS NULL;
CREATE INDEX idx_tickets_assignee ON tickets(assignee_id, status)
    WHERE deleted_at IS NULL AND assignee_id IS NOT NULL;
CREATE INDEX idx_tickets_search ON tickets USING GIN(search_vector);
CREATE INDEX idx_tickets_created ON tickets(tenant_id, created_at DESC)
    WHERE deleted_at IS NULL;

-- RLS Policy
ALTER TABLE tickets ENABLE ROW LEVEL SECURITY;
CREATE POLICY tenant_isolation ON tickets
    USING (tenant_id = current_setting('app.current_tenant_id')::UUID);
```

Query Optimization

Analyzing Slow Queries

Feed AI your slow query and its EXPLAIN output:

```
This query takes 4.2 seconds in production. Help me optimize it.

Query:
SELECT t.*,
       COUNT(c.id) as comment_count,
       MAX(c.created_at) as last_comment_at
FROM tickets t
LEFT JOIN comments c ON c.ticket_id = t.id
WHERE t.tenant_id = '...'
  AND t.status IN ('open', 'in_progress')
  AND t.created_at > NOW() - INTERVAL '90 days'
```

```
GROUP BY t.id
ORDER BY t.priority DESC, t.created_at DESC
LIMIT 50;

EXPLAIN ANALYZE output:
Sort  (cost=45231..45232 rows=50 width=324) (actual time=4187..4188 rows=50 loop
s=1)
  Sort Key: t.priority DESC, t.created_at DESC
  ->  HashAggregate  (cost=45100..45200 rows=2500 width=324) (actual time=4150..
4170 rows=2500 loops=1)
        ->  Hash Left Join  (cost=35000..42000 rows=125000 width=290) (actual ti
me=2100..3900 rows=125000 loops=1)
              ->  Seq Scan on tickets t  (cost=0..15000 rows=2500 width=280) (ac
tual time=0.1..800 rows=2500 loops=1)
                    Filter: (tenant_id = '...' AND status IN ('open','in_progres
s') AND created_at > ...)
                    Rows Removed by Filter: 497500
              ->  Hash  (cost=25000..25000 rows=500000 width=30) (actual time=12
00..1200 rows=500000 loops=1)
                    ->  Seq Scan on comments c  (cost=0..25000 rows=500000 width
=30) (actual time=0.1..900 rows=500000 loops=1)
Planning Time: 0.5 ms
Execution Time: 4192 ms
```

AI's analysis:

```
Problems identified:

1. Sequential scan on tickets (500K rows scanned, 2.5K returned)
   → Missing composite index on (tenant_id, status, created_at)

2. Sequential scan on comments (500K rows scanned for the join)
   → Missing index on comments(ticket_id)

3. JOIN inflates rows before aggregation
   → Consider a lateral join or subquery for counts

Recommended fixes:

-- Index for the ticket filter
CREATE INDEX idx_tickets_tenant_status_created
  ON tickets(tenant_id, status, created_at DESC)
  WHERE deleted_at IS NULL;

-- Index for the comment join
```

```
CREATE INDEX idx_comments_ticket_id
  ON comments(ticket_id);

-- Rewritten query using lateral join (avoids GROUP BY inflation):
SELECT t.*,
       c.comment_count,
       c.last_comment_at
FROM tickets t
CROSS JOIN LATERAL (
    SELECT COUNT(*) as comment_count,
           MAX(created_at) as last_comment_at
    FROM comments
    WHERE ticket_id = t.id
) c
WHERE t.tenant_id = '...'
  AND t.status IN ('open', 'in_progress')
  AND t.created_at > NOW() - INTERVAL '90 days'
  AND t.deleted_at IS NULL
ORDER BY t.priority DESC, t.created_at DESC
LIMIT 50;

Expected improvement: 4200ms → <50ms
```

Migration Generation: A Practical Example

Before we discuss migration management in the abstract, here's a real workflow. You need to add a "teams" feature to an existing application. Users can belong to multiple teams, and each team has an owner.

Your prompt:

```
Generate a Prisma migration and updated schema for adding teams.
Current schema has a User model with id, email, name, createdAt.

New requirements:
- Team model: id, name, slug (unique), createdAt
- TeamMember junction: userId, teamId, role (owner/admin/member), joinedAt
- A user can belong to multiple teams
- Each team must have exactly one owner
- Slug must be URL-safe (lowercase alphanumeric and hyphens)
- Add indexes for common queries: list teams for a user, list members for a team
```

AI generates the Prisma schema update:

```
model Team {
  id        String      @id @default(cuid())
  name      String
  slug      String      @unique
  createdAt DateTime    @default(now())
  members   TeamMember[]

  @@index([slug])
}

model TeamMember {
  userId   String
  teamId   String
  role     TeamRole   @default(MEMBER)
  joinedAt DateTime   @default(now())
  user     User       @relation(fields: [userId], references: [id], onDelete: Ca
scade)
  team     Team       @relation(fields: [teamId], references: [id], onDelete: Ca
scade)

  @@id([userId, teamId])
  @@index([teamId, role])
  @@index([userId])
}

enum TeamRole {
  OWNER
  ADMIN
  MEMBER
}
```

You run `npx prisma migrate dev --name add-teams` and Prisma generates the SQL migration. But you also need a data migration — existing users may need to land in a default team. This is where AI helps with the follow-up:

```
Now generate a data migration script that:
1. Creates a "Personal" team for every existing user
2. Adds each user as the OWNER of their personal team
3. Generates a slug from the user's name (lowercase, hyphenated)
4. Handles slug collisions by appending a random suffix
5. Runs in batches of 500 to avoid memory issues
6. Is idempotent (safe to run multiple times)
```

AI generates:

```
import { PrismaClient } from '@prisma/client';
import { createId } from '@paralleldrive/cuid2';

const prisma = new PrismaClient();
const BATCH_SIZE = 500;

function slugify(name: string): string {
  return name.toLowerCase().replace(/[^a-z0-9]+/g, '-').replace(/(^-|-$)/g, '');
}

async function main() {
  let skip = 0;

  while (true) {
    const users = await prisma.user.findMany({
      where: {
        teams: { none: {} },  // Only users without any team
      },
      take: BATCH_SIZE,
      skip,
    });

    if (users.length === 0) break;

    for (const user of users) {
      let slug = slugify(user.name) || 'user';
      const existing = await prisma.team.findUnique({ where: { slug } });
      if (existing) slug = `${slug}-${createId().slice(0, 6)}`;

      await prisma.team.create({
        data: {
          name: `${user.name}'s Team`,
          slug,
          members: {
            create: { userId: user.id, role: 'OWNER' },
          },
        },
      });
    }

    skip += BATCH_SIZE;
    console.log(`Processed ${skip} users...`);
  }
}
```

```
main().catch(console.error).finally(() => prisma.$disconnect());
```

This is a strong migration script. The idempotency check (only processing users without teams) lets you safely re-run it if it fails partway through. Batching prevents memory issues on large user tables.

What to verify: Check the `skip` variable against the `where` filter. The `where` clause excludes already-migrated users, so `skip` should stay at 0 each iteration — the result set shrinks as users get teams. The current code increments `skip`, which could skip users entirely. It's a subtle bug AI introduces because the batching pattern from its training data does not account for the narrowing `where` clause. Fix it by removing the `skip` parameter and always taking the first batch of unprocessed users.

Migration Management

Generating Migrations

```
claude "Generate a database migration to add the following changes:
  1. Add a 'priority' column to the tickets table (TEXT, default 'medium')
  2. Add a 'tags' table (id, tenant_id, name, color)
  3. Add a 'ticket_tags' junction table
  4. Add an index on tickets(priority)

  Use our migration tool (Prisma Migrate).
  Include both the up and down migrations.
  Make the migration safe for zero-downtime deployment."
```

Zero-Downtime Migration Strategies

```
I need to rename a column from 'user_name' to 'display_name'
in the users table. This table has 2M rows and the column is
referenced by 5 microservices. We cannot have downtime.

Design a migration strategy that:
1. Maintains backward compatibility during rollout
2. Can be rolled back at any step
3. Does not lock the table for extended periods
```

AI outlines a multi-step migration:

```sql
-- Step 1: Add new column (no lock, fast)
ALTER TABLE users ADD COLUMN display_name TEXT;

-- Step 2: Backfill (batched, no lock)
-- Note: PostgreSQL does not support UPDATE ... LIMIT.
-- Use a subquery to batch the updates.
UPDATE users SET display_name = user_name
WHERE id IN (
    SELECT id FROM users WHERE display_name IS NULL LIMIT 10000
);  -- Run in batches until no rows remain

-- Step 3: Add trigger to keep columns in sync
CREATE OR REPLACE FUNCTION sync_display_name()
RETURNS TRIGGER AS $$
BEGIN
    IF NEW.user_name IS DISTINCT FROM OLD.user_name THEN
        NEW.display_name := NEW.user_name;
    END IF;
    IF NEW.display_name IS DISTINCT FROM OLD.display_name THEN
        NEW.user_name := NEW.display_name;
    END IF;
    RETURN NEW;
END;
$$ LANGUAGE plpgsql;

-- Step 4: Update services one by one to use display_name

-- Step 5: After all services migrated, drop old column
-- ALTER TABLE users DROP COLUMN user_name;
```

Query Generation

Complex Reporting Queries

```
Write a PostgreSQL query for a sales dashboard that shows:
- Monthly revenue for the last 12 months
- Comparison with same month previous year
- Month-over-month growth percentage
- Running total (cumulative revenue)
- Top 5 products by revenue per month

The relevant tables are:
- orders (id, user_id, status, total, created_at)
- order_items (id, order_id, product_id, quantity, unit_price)
```

```
- products (id, name, category)

Only include orders with status = 'completed'.

-- Monthly revenue with YoY comparison
WITH monthly_revenue AS (
    SELECT
        date_trunc('month', o.created_at) AS month,
        SUM(o.total) AS revenue,
        COUNT(DISTINCT o.id) AS order_count,
        COUNT(DISTINCT o.user_id) AS unique_customers
    FROM orders o
    WHERE o.status = 'completed'
      AND o.created_at >= date_trunc('month', NOW()) - INTERVAL '12 months'
    GROUP BY date_trunc('month', o.created_at)
),
previous_year AS (
    SELECT
        date_trunc('month', o.created_at) + INTERVAL '12 months' AS month,
        SUM(o.total) AS revenue_prev_year
    FROM orders o
    WHERE o.status = 'completed'
      AND o.created_at >= date_trunc('month', NOW()) - INTERVAL '24 months'
      AND o.created_at < date_trunc('month', NOW()) - INTERVAL '12 months'
    GROUP BY date_trunc('month', o.created_at)
)
SELECT
    mr.month,
    mr.revenue,
    mr.order_count,
    mr.unique_customers,
    py.revenue_prev_year,
    CASE
        WHEN py.revenue_prev_year > 0
        THEN ROUND((mr.revenue - py.revenue_prev_year) / py.revenue_prev_year *
100, 1)
        ELSE NULL
    END AS yoy_growth_pct,
    ROUND(
        (mr.revenue - LAG(mr.revenue) OVER (ORDER BY mr.month)) /
        NULLIF(LAG(mr.revenue) OVER (ORDER BY mr.month), 0) * 100, 1
    ) AS mom_growth_pct,
    SUM(mr.revenue) OVER (ORDER BY mr.month) AS cumulative_revenue
FROM monthly_revenue mr
```

```sql
LEFT JOIN previous_year py ON py.month = mr.month
ORDER BY mr.month;
```

ORM Best Practices

AI helps you use ORMs correctly, avoiding common pitfalls:

```
claude "Review our Prisma queries in src/repositories/ for:
  1. N+1 query patterns (missing includes/joins)
  2. Over-fetching (selecting all columns when we need only a few)
  3. Missing transactions where multiple writes should be atomic
  4. Raw queries that could use the ORM
  5. ORM queries that would be better as raw SQL (complex aggregations)"
```

What AI Gets Wrong with Databases

Missing indexes for new queries: AI writes queries without checking whether the required indexes exist.

Unsafe migrations: AI generates ALTER TABLE statements that lock tables for long stretches on large datasets.

Incorrect transaction isolation: AI defaults to the database's default isolation level, which may not fit your use case.

Over-normalized schemas: AI sometimes normalizes too aggressively, forcing excessive JOINs for common queries.

Ignoring data volume: AI generates queries that work on 1,000 rows but fail on 10 million. Always state your data volume.

Test AI-generated queries against realistic data volumes, and test migrations on a copy of production data before running them in production.

One more pattern worth mentioning: AI excels at generating seed data for development and testing. If you need realistic test data for a schema with complex relationships, describe the schema and constraints and ask AI to generate INSERT statements. It handles foreign key ordering, enum constraints, and realistic-looking data (proper names, valid email formats, sensible dates) far better than handwritten test fixtures. For large volumes, ask for a script that produces N rows in a loop rather than individual INSERT statements.

From Server-Side Storage to Pocket-Sized Apps

Databases power backends. Mobile apps consume those backends. Chapter 19 covers AI-assisted mobile development — building for iOS, Android, and cross-platform with AI tools that understand mobile-specific patterns and constraints.

Chapter 19: Mobile Development with AI

iOS, Android, Cross-Platform with AI Assistance

Mobile development has unique challenges that AI handles with varying degrees of success. Platform-specific APIs, UI guidelines, performance constraints, and app store requirements create a specialized domain where generic AI advice falls short but targeted AI assistance produces excellent results.

This chapter covers using AI for iOS (Swift/SwiftUI), Android (Kotlin/Jetpack Compose), and cross-platform (React Native, Flutter) development — what works well, what needs careful verification, and which tools fit each platform.

iOS Development with AI

SwiftUI Component Generation

AI generates SwiftUI code effectively because SwiftUI's declarative syntax maps cleanly to natural language descriptions:

```
Create a SwiftUI view for a product detail screen:
- Hero image at top (full width, aspect ratio preserved)
- Product name (large title)
- Price with strikethrough original if discounted
- Star rating display (1-5 stars, supports half stars)
- Description text (expandable, truncated to 3 lines by default)
- "Add to Cart" button (bottom, full width, disabled if out of stock)
- Supports dark mode
- Use proper accessibility labels
```

```swift
import SwiftUI

struct ProductDetailView: View {
    let product: Product
    @State private var isDescriptionExpanded = false
    @State private var addingToCart = false

    var body: some View {
```

```
        ScrollView {
            VStack(alignment: .leading, spacing: 16) {
                // Hero image
                AsyncImage(url: product.imageURL) { image in
                    image
                        .resizable()
                        .aspectRatio(contentMode: .fill)
                } placeholder: {
                    Rectangle()
                        .fill(Color.gray.opacity(0.2))
                        .overlay(ProgressView())
                }
                .frame(height: 300)
                .clipped()

                VStack(alignment: .leading, spacing: 12) {
                    // Product name
                    Text(product.name)
                        .font(.largeTitle)
                        .fontWeight(.bold)

                    // Price
                    HStack(spacing: 8) {
                        if let discountedPrice = product.discountedPrice {
                            Text(discountedPrice.formatted(.currency(code: "USD
")))
                                .font(.title2)
                                .fontWeight(.semibold)
                                .foregroundColor(.green)

                            Text(product.price.formatted(.currency(code: "USD")))
                                .font(.body)
                                .strikethrough()
                                .foregroundColor(.secondary)
                        } else {
                            Text(product.price.formatted(.currency(code: "USD")))
                                .font(.title2)
                                .fontWeight(.semibold)
                        }
                    }
                    .accessibilityElement(children: .combine)
                    .accessibilityLabel(priceAccessibilityLabel)

                    // Star rating
                    StarRatingView(rating: product.rating, maxRating: 5)
```

```
                        .accessibilityLabel("\(product.rating, specifier: "%.1f")
out of 5 stars")

                    // Description
                    VStack(alignment: .leading, spacing: 4) {
                        Text(product.description)
                            .lineLimit(isDescriptionExpanded ? nil : 3)
                            .animation(.easeInOut, value: isDescriptionExpanded)

                        Button(isDescriptionExpanded ? "Show less" : "Read more")
{
                            isDescriptionExpanded.toggle()
                        }
                        .font(.caption)
                    }

                    // Stock status
                    if !product.isInStock {
                        Label("Out of Stock", systemImage: "exclamationmark.tria
ngle")
                            .foregroundColor(.red)
                            .font(.callout)
                    }
                }
                .padding(.horizontal)
            }
        }
        .safeAreaInset(edge: .bottom) {
            Button(action: addToCart) {
                HStack {
                    if addingToCart {
                        ProgressView()
                            .tint(.white)
                    } else {
                        Image(systemName: "cart.badge.plus")
                        Text("Add to Cart")
                    }
                }
                .frame(maxWidth: .infinity)
                .padding()
                .background(product.isInStock ? Color.accentColor : Color.gray)
                .foregroundColor(.white)
                .cornerRadius(12)
            }
            .disabled(!product.isInStock || addingToCart)
```

```
            .padding()
            .background(.bar)
        }
    }

    private var priceAccessibilityLabel: String {
        if let discountedPrice = product.discountedPrice {
            return "On sale for \(discountedPrice.formatted(.currency(code: "USD
"))), was \(product.price.formatted(.currency(code: "USD")))"
        }
        return product.price.formatted(.currency(code: "USD"))
    }

    private func addToCart() {
        addingToCart = true
        // Cart logic
    }
}

struct StarRatingView: View {
    let rating: Double
    let maxRating: Int

    var body: some View {
        HStack(spacing: 4) {
            ForEach(0..<maxRating, id: \.self) { index in
                Image(systemName: starImage(for: index))
                    .foregroundColor(.yellow)
                    .font(.caption)
            }
        }
    }

    private func starImage(for index: Int) -> String {
        let threshold = Double(index)
        if rating >= threshold + 1 {
            return "star.fill"
        } else if rating >= threshold + 0.5 {
            return "star.leadinghalf.filled"
        } else {
            return "star"
        }
    }
}
```

iOS-Specific Patterns

AI handles iOS patterns well when you are specific:

```
Implement a network layer for an iOS app using async/await:
- URLSession-based with custom configuration
- Automatic retry with exponential backoff (3 attempts)
- Authentication token injection via request interceptor
- Token refresh on 401 responses
- Request/response logging in debug builds
- Codable model decoding with custom date format
- Error handling with typed errors
- Cancel support via Task cancellation
```

Android Development with AI

Jetpack Compose

```
Create an Android composable for a settings screen:
- Grouped sections (Account, Notifications, Privacy, About)
- Toggle switches for boolean settings
- Navigation items (arrow icon, navigate to detail)
- Current value display for selection items
- Material 3 design system
- Dark mode support
- Preview annotations
```

```kotlin
@OptIn(ExperimentalMaterial3Api::class)
@Composable
fun SettingsScreen(
    viewModel: SettingsViewModel = hiltViewModel(),
    onNavigate: (SettingsDestination) -> Unit
) {
    val uiState by viewModel.uiState.collectAsStateWithLifecycle()

    Scaffold(
        topBar = {
            TopAppBar(title = { Text("Settings") })
        }
    ) { padding ->
        LazyColumn(
            modifier = Modifier
                .fillMaxSize()
                .padding(padding)
```

```
) {
    // Account Section
    item {
        SettingsSectionHeader(title = "Account")
    }
    item {
        SettingsNavigationItem(
            title = "Profile",
            subtitle = uiState.userEmail,
            icon = Icons.Default.Person,
            onClick = { onNavigate(SettingsDestination.Profile) }
        )
    }
    item {
        SettingsNavigationItem(
            title = "Subscription",
            subtitle = uiState.planName,
            icon = Icons.Default.Star,
            onClick = { onNavigate(SettingsDestination.Subscription) }
        )
    }

    // Notifications Section
    item {
        SettingsSectionHeader(title = "Notifications")
    }
    item {
        SettingsToggleItem(
            title = "Push Notifications",
            subtitle = "Receive alerts on your device",
            icon = Icons.Default.Notifications,
            checked = uiState.pushEnabled,
            onCheckedChange = viewModel::setPushEnabled
        )
    }
    item {
        SettingsToggleItem(
            title = "Email Notifications",
            subtitle = "Weekly digest and updates",
            icon = Icons.Default.Email,
            checked = uiState.emailEnabled,
            onCheckedChange = viewModel::setEmailEnabled
        )
    }
```

```
            // About Section
            item {
                SettingsSectionHeader(title = "About")
            }
            item {
                SettingsNavigationItem(
                    title = "Version",
                    subtitle = BuildConfig.VERSION_NAME,
                    icon = Icons.Default.Info,
                    onClick = { }
                )
            }
        }
    }
}

@Composable
private fun SettingsSectionHeader(title: String) {
    Text(
        text = title,
        style = MaterialTheme.typography.titleSmall,
        color = MaterialTheme.colorScheme.primary,
        modifier = Modifier.padding(horizontal = 16.dp, vertical = 8.dp)
    )
}

@Composable
private fun SettingsToggleItem(
    title: String,
    subtitle: String,
    icon: ImageVector,
    checked: Boolean,
    onCheckedChange: (Boolean) -> Unit
) {
    ListItem(
        headlineContent = { Text(title) },
        supportingContent = { Text(subtitle) },
        leadingContent = {
            Icon(icon, contentDescription = null, tint = MaterialTheme.colorSche
me.onSurfaceVariant)
        },
        trailingContent = {
            Switch(checked = checked, onCheckedChange = onCheckedChange)
        },
        modifier = Modifier.clickable { onCheckedChange(!checked) }
```

```
    )
}

@Preview(showBackground = true)
@Preview(showBackground = true, uiMode = Configuration.UI_MODE_NIGHT_YES)
@Composable
private fun SettingsScreenPreview() {
    AppTheme {
        SettingsScreen(onNavigate = {})
    }
}
```

Android-Specific Tools

Gemini Code Assist excels here. Google's AI tool has deep knowledge of Android APIs, Kotlin patterns, and Jetpack libraries. If you do Android-first development, Gemini is worth adding to your toolkit alongside your general-purpose AI tool.

```
## Gemini-specific prompt for Android
Create a ViewModel for the settings screen using:
- Hilt dependency injection
- StateFlow for UI state
- DataStore Preferences for persistence
- Proper lifecycle handling (viewModelScope)
- Error handling with sealed class UiEvent
```

Cross-Platform Development

React Native

```
Create a React Native component for an image carousel:
- Horizontal scrolling with snap-to-item
- Page indicator dots
- Auto-play with pause on touch
- Supports both iOS and Android gestures
- Uses react-native-reanimated for smooth animations
- TypeScript
- Handles image loading states
```

AI generates platform-aware code:

```
import React, { useCallback, useEffect, useRef, useState } from 'react';
import { Dimensions, FlatList, Image, StyleSheet, View } from 'react-native';
```

```
import Animated, {
  useAnimatedStyle,
  withTiming,
  interpolateColor
} from 'react-native-reanimated';

const { width: SCREEN_WIDTH } = Dimensions.get('window');

interface CarouselProps {
  images: { uri: string; alt: string }[];
  autoPlayInterval?: number;
  height?: number;
}

export function ImageCarousel({
  images,
  autoPlayInterval = 4000,
  height = 250
}: CarouselProps) {
  const [activeIndex, setActiveIndex] = useState(0);
  const flatListRef = useRef<FlatList>(null);
  const autoPlayRef = useRef<ReturnType<typeof setInterval>>();
  const isPausedRef = useRef(false);

  const startAutoPlay = useCallback(() => {
    autoPlayRef.current = setInterval(() => {
      if (!isPausedRef.current) {
        setActiveIndex((prev) => {
          const next = (prev + 1) % images.length;
          flatListRef.current?.scrollToIndex({ index: next, animated: true });
          return next;
        });
      }
    }, autoPlayInterval);
  }, [images.length, autoPlayInterval]);

  useEffect(() => {
    if (images.length > 1) {
      startAutoPlay();
    }
    return () => {
      if (autoPlayRef.current) clearInterval(autoPlayRef.current);
    };
  }, [startAutoPlay, images.length]);
```

```
  const onViewableItemsChanged = useRef(({ viewableItems }: any) => {
    if (viewableItems.length > 0) {
      setActiveIndex(viewableItems[0].index ?? 0);
    }
  }).current;

  return (
    <View style={[styles.container, { height }]}>
      <FlatList
        ref={flatListRef}
        data={images}
        horizontal
        pagingEnabled
        showsHorizontalScrollIndicator={false}
        onViewableItemsChanged={onViewableItemsChanged}
        viewabilityConfig={{ itemVisiblePercentThreshold: 50 }}
        onTouchStart={() => { isPausedRef.current = true; }}
        onTouchEnd={() => { isPausedRef.current = false; }}
        renderItem={({ item }) => (
          <Image
            source={{ uri: item.uri }}
            style={[styles.image, { height }]}
            resizeMode="cover"
            accessibilityLabel={item.alt}
          />
        )}
        keyExtractor={(_, index) => index.toString()}
      />
      {images.length > 1 && (
        <View style={styles.pagination}>
          {images.map((_, index) => (
            <View
              key={index}
              style={[
                styles.dot,
                index === activeIndex && styles.activeDot,
              ]}
            />
          ))}
        </View>
      )}
    </View>
  );
}
```

```
const styles = StyleSheet.create({
  container: { position: 'relative' },
  image: { width: SCREEN_WIDTH },
  pagination: {
    position: 'absolute',
    bottom: 12,
    flexDirection: 'row',
    alignSelf: 'center',
    gap: 8,
  },
  dot: {
    width: 8,
    height: 8,
    borderRadius: 4,
    backgroundColor: 'rgba(255, 255, 255, 0.5)',
  },
  activeDot: {
    backgroundColor: '#fff',
    width: 24,
  },
});
```

Flutter

```
// Prompt: "Create a Flutter widget for a bottom sheet with a form"

class AddTaskBottomSheet extends StatefulWidget {
  final Function(TaskData) onSubmit;

  const AddTaskBottomSheet({super.key, required this.onSubmit});

  @override
  State<AddTaskBottomSheet> createState() => _AddTaskBottomSheetState();
}

class _AddTaskBottomSheetState extends State<AddTaskBottomSheet> {
  final _formKey = GlobalKey<FormState>();
  final _titleController = TextEditingController();
  final _descriptionController = TextEditingController();
  DateTime? _dueDate;
  Priority _priority = Priority.medium;
  bool _isSubmitting = false;

  @override
  void dispose() {
```

```
    _titleController.dispose();
    _descriptionController.dispose();
    super.dispose();
  }

  @override
  Widget build(BuildContext context) {
    return Padding(
      padding: EdgeInsets.only(
        bottom: MediaQuery.of(context).viewInsets.bottom,
        left: 16, right: 16, top: 16,
      ),
      child: Form(
        key: _formKey,
        child: Column(
          mainAxisSize: MainAxisSize.min,
          crossAxisAlignment: CrossAxisAlignment.stretch,
          children: [
            // Drag handle
            Center(
              child: Container(
                width: 40, height: 4,
                decoration: BoxDecoration(
                  color: Colors.grey[300],
                  borderRadius: BorderRadius.circular(2),
                ),
              ),
            ),
            const SizedBox(height: 16),
            Text('New Task', style: Theme.of(context).textTheme.headlineSmall),
            const SizedBox(height: 16),
            TextFormField(
              controller: _titleController,
              decoration: const InputDecoration(
                labelText: 'Title',
                border: OutlineInputBorder(),
              ),
              validator: (value) =>
                  value?.isEmpty ?? true ? 'Title is required' : null,
              autofocus: true,
            ),
            const SizedBox(height: 12),
            TextFormField(
              controller: _descriptionController,
              decoration: const InputDecoration(
```

```
              labelText: 'Description (optional)',
              border: OutlineInputBorder(),
            ),
            maxLines: 3,
          ),
          const SizedBox(height: 16),
          FilledButton(
            onPressed: _isSubmitting ? null : _submit,
            child: _isSubmitting
                ? const SizedBox(
                    height: 20, width: 20,
                    child: CircularProgressIndicator(strokeWidth: 2),
                  )
                : const Text('Create Task'),
          ),
          const SizedBox(height: 16),
        ],
      ),
    ),
  );
}

Future<void> _submit() async {
  if (!_formKey.currentState!.validate()) return;
  setState(() => _isSubmitting = true);
  // Submit logic
}
}
```

Cross-Platform Workflow: Building a Feature in React Native

Here is a realistic cross-platform workflow that shows how AI handles the platform-specific nuances that make mobile development tricky.

The task: Build a "pull to refresh" list screen that fetches data from an API, works on both iOS and Android, handles loading and error states, and persists the last fetched data for offline access.

Step 1: Generate the screen with platform considerations.

```
Create a React Native screen for a notifications list:
- Pull to refresh
```

```
- Infinite scroll (load more on reaching bottom)
- Each notification has: icon, title, body, timestamp, read/unread state
- Swipe to dismiss (react-native-gesture-handler)
- Persist last fetched data with AsyncStorage for offline display
- Show "No connection" banner when offline (use @react-native-community/netinfo)
- TypeScript, functional component
```

AI generates a component that handles the happy path well. But here is where platform differences matter.

Step 2: Fix platform-specific issues.

```
Two platform issues:
1. On Android, the pull-to-refresh spinner color should use our
   brand color (the default is the system green)
2. On iOS, the FlatList bounce is hiding the refresh indicator.
   Set the progressViewOffset for Android.

Also, the swipe-to-dismiss gesture is conflicting with the
navigation drawer gesture on Android. Use a horizontal swipe
threshold of 40px to prevent accidental triggers.
```

You only learn this kind of thing from testing on actual devices. AI knows about `refreshControl` color props and gesture thresholds because it has seen these fixes in thousands of projects, but it does not apply them proactively.

Step 3: Add offline persistence properly.

AI's first version of offline persistence usually has a flaw — it reads from AsyncStorage on every render instead of initializing from cache and then updating. You refine:

```
The offline caching should work like this:
1. On mount, immediately show cached data from AsyncStorage (stale-while-revalid
ate)
2. Simultaneously fetch fresh data from the API
3. If the fetch succeeds, update the display AND the cache
4. If the fetch fails and we have cached data, show cached data with a
   "Last updated: 2 hours ago" indicator
5. If the fetch fails and we have NO cached data, show an error with retry butto
n
```

Stale-while-revalidate is the correct pattern for mobile apps, and AI implements it well once you describe the behavior explicitly. You just need to ask — AI's default is fetch-or-fail, which makes for a poor mobile experience.

Flutter equivalent: The same workflow applies in Flutter, but the tools differ. Instead of AsyncStorage, you use `shared_preferences` or `hive` for caching. Instead of `react-native-gesture-handler`, you use Flutter's built-in `Dismissible` widget. The prompting strategy is the same: generate the base, test on both platforms, fix platform-specific issues, refine the offline behavior.

What AI Gets Wrong in Mobile

Platform guidelines: AI might generate iOS patterns in Android code or vice versa. Specify the platform explicitly.

Lifecycle handling: Mobile apps have complex lifecycles. AI sometimes skips lifecycle-appropriate cleanup (disposing controllers, cancelling subscriptions).

Memory management: Mobile apps run in constrained environments. AI may generate code that loads all data into memory, which crashes on devices with limited RAM.

Permissions: AI often forgets to request runtime permissions before accessing camera, location, or storage.

Offline support: AI generates code that assumes network availability. Real mobile apps need offline-first patterns.

Always test AI-generated mobile code on actual devices, not just simulators. Real-world conditions — slow networks, low memory, background mode — surface issues that simulators hide.

App store compliance: AI has no awareness of Apple's App Store Review Guidelines or Google Play policies. It may generate code that uses private APIs (instant rejection on iOS), requests unnecessary permissions (flagged by Google Play), or implements in-app purchases without using the platform's required payment system. Always cross-reference AI-generated features against the current platform guidelines before submission.

Mobile, web, backend, infrastructure — that's AI assistance across every development domain. Chapter 20 shifts from what AI helps you build to how to work with it effectively. AI pair programming is about making AI your ongoing collaborator, not just a tool you reach for on isolated tasks.

Part VI: Working with AI

"The best programs are written so that computing machines can perform them quickly and so that human beings can understand them clearly."

— Donald Knuth

Chapter 20: The AI Pair Programming Workflow

How to Actually Work WITH AI, Not Just Ask It Questions

There is a difference between using AI and working with AI. Most developers use AI as a question-answering service: ask a question, get an answer, go back to coding alone. That leaves most of the value on the table.

Working with AI means maintaining an ongoing collaborative relationship throughout a development session. You think out loud. AI fills in details. You catch mistakes. AI catches yours. The result is qualitatively different from either working alone or asking occasional questions — it is actual pair programming, with a partner that never tires, never gets defensive about its code, and always brings a fresh perspective.

This chapter teaches you how to build that workflow.

The Pair Programming Mental Model

In traditional pair programming, there is a driver (who types) and a navigator (who thinks ahead, catches errors, and watches the bigger picture). With AI pair programming, the roles are more fluid:

You are always the navigator. You set direction, make design decisions, and hold the vision of what you are building.

AI is usually the driver. It writes the code, suggests implementations, and handles mechanical details.

But you switch roles constantly. Sometimes you type and AI reviews. Sometimes AI generates an approach and you critique it. Sometimes you both contribute ideas and the conversation produces something neither would have built alone.

The key insight: the value is in the conversation, not in any single AI output.

The Session Workflow

Phase 1: Context Setting (2–5 minutes)

Every productive AI pair programming session starts with context. Don't jump straight into "write me a function." Set the scene:

```
I'm working on the order fulfillment module today. Here's what
I need to accomplish:

1. Add a background job that checks for orders stuck in
   "processing" for more than 30 minutes
2. The job should retry the payment processing and, if that fails,
   cancel the order and notify the customer
3. We need to handle the case where the payment actually went
   through but we lost the confirmation

Our stack: Node.js, TypeScript, BullMQ for jobs, Stripe for payments,
and Prisma for database. The relevant code is in
src/modules/orders/ and src/jobs/.

Let's start by discussing the approach before writing any code.
```

This context setting takes two minutes and saves 30. AI now knows what you are building, why, and how it fits into your system.

Phase 2: Design Discussion (5–10 minutes)

Before writing code, talk through the design:

```
For the stuck order detection, I'm thinking:
1. A cron job that runs every 5 minutes
2. Queries for orders in 'processing' status older than 30 minutes
3. For each, attempts to verify payment status with Stripe
4. Three outcomes: payment succeeded (update order), payment failed
   (cancel and notify), payment unknown (retry later)

What am I missing? What failure modes should I consider?
```

AI might surface considerations you had not thought of: - What if the cron job itself fails mid-processing? (Idempotency needed) - What if Stripe is down when you check payment status? (Retry vs. skip) - What if multiple job instances run at once? (Distributed lock) - Should there be a maximum retry count before human intervention?

This conversation refines your design before any code exists.

Phase 3: Implementation in Chunks (30–60 minutes)

Break the work into small, reviewable pieces:

```
Let's start with the stuck order query. Write a function that:
- Finds orders in 'processing' status older than 30 minutes
- Excludes orders that already have a retry job scheduled
- Returns them sorted by oldest first
- Limits to 100 per batch (to avoid overwhelming Stripe)
```

Review the output. If it is good, move on:

```
Good. Now let's add the Stripe verification. For each stuck order,
we need to check the payment intent status. Write a function that:
- Takes an order with its Stripe payment intent ID
- Calls Stripe to get the current payment intent status
- Maps the Stripe status to our internal resolution:
  - succeeded → confirm order
  - canceled/failed → cancel order
  - requires_action/processing → retry later
- Handles Stripe API errors gracefully
```

Each chunk gets reviewed, tested, and committed before you move to the next.

Phase 4: Review and Polish (10–15 minutes)

After the feature is implemented:

```
We've got the basic implementation working. Now let's review
the whole thing:
1. Are there any edge cases we missed?
2. Does the error handling cover all failure modes?
3. Are the log messages useful for debugging production issues?
4. Should we add any metrics or alerts?
5. Is the code consistent with our existing patterns?
```

This review phase often catches subtle issues the per-chunk reviews missed — especially interactions between the chunks.

Communication Patterns That Work

The Thinking Out Loud Pattern

Share your reasoning, not just your requirements:

```
I'm torn between using a database flag (is_stuck_checked) and
a separate tracking table (stuck_order_checks) for tracking
which orders have been processed by the cron job.

The flag is simpler but adds a column to the orders table.
The tracking table is more complex but keeps the orders table clean
and gives us an audit trail of all check attempts.

What would you recommend given that we might need to audit
these checks for compliance?
```

AI gives much better advice when it understands your reasoning and constraints.

The Rubber Duck Pattern

Sometimes you don't need AI to solve your problem. You need to explain your problem clearly enough that the solution becomes obvious:

```
I'm debugging something weird. The cron job processes orders
correctly on the first run, but on the second run, it picks up
the same orders again even though they've been resolved.

Let me trace through the code...

The query filters by status = 'processing'. After resolution,
we update the status to 'completed' or 'cancelled'. So why
are they showing up again?

Wait – I think the status update is happening inside a transaction
that hasn't committed by the time the next job runs. The job
interval is 5 minutes but the transaction can take up to 3 minutes
for large batches. Is that possible?
```

You just solved your own problem. But AI might add: "Yes, and consider using SELECT FOR UPDATE SKIP LOCKED so the next job run skips orders that are currently being processed by the previous run."

The Incremental Specification Pattern

Start vague, get specific through conversation:

```
You: I need error handling for the Stripe verification.

AI: What kinds of errors? Network errors, authentication errors,
    rate limits, invalid payment intent IDs?

You: All of those. Network errors should retry. Auth errors
     should alert the team. Rate limits should back off.
     Invalid IDs should log and skip.

AI: Should retries happen within this job run or be scheduled
    as a separate job? What's the maximum retry count?

You: Separate job, max 3 retries, then alert for manual review.

AI: [generates comprehensive error handling based on the
    refined specification]
```

Each exchange narrows the specification until AI has enough detail to produce strong output.

Multi-Tool Pair Programming

The best pair programming sessions use multiple tools:

The Chat + Editor Workflow

1. **Open a chat session** (Claude web, ChatGPT) for design discussion
2. **Use Claude Code or Cursor** for implementation
3. **Use Copilot** for small inline completions
4. **Return to chat** when you hit a decision point

```
Chat: "What's the best way to handle partial failures in a batch job?"
[Discussion about patterns]

Editor: "Implement the batch processor with the circuit breaker
  pattern we discussed"
[Implementation]

Chat: "The implementation is working but I'm seeing memory growth
```

```
  over time. Here's the heap snapshot..."
[Debugging conversation]
```

The Conversation Continuity Advantage

Unlike one-off prompts, pair programming sessions build context over time. An AI that has been working with you for an hour knows:

- Your codebase structure
- Your design preferences
- Decisions you have already made
- Mistakes you have already corrected
- The specific problem you are solving

This accumulated context produces dramatically better output than a cold start. Protect long-running sessions — save context, keep conversations focused, avoid switching topics mid-session.

A Real Session Transcript

Theory is useful, but seeing the actual back-and-forth is more useful. Here is an abbreviated transcript from a real pair programming session building a rate limiter middleware. The conversation is condensed but the rhythm is preserved.

```
You:   I need a rate limiter for our API. We're using Redis and Express.
       Before writing code, let's discuss the algorithm. I've heard of
       token bucket, sliding window, and fixed window. Which fits best
       for an API that has per-user limits of 100 requests per minute?

AI:    For per-user API rate limiting, sliding window log or sliding
       window counter are the best fits. Here's why:
       - Fixed window has the burst problem at window boundaries (a user
         can make 100 requests at 11:59:59 and 100 more at 12:00:00)
       - Token bucket is great for smoothing but more complex to implement
       - Sliding window counter gives you accurate-enough counting without
         storing every request timestamp

       For 100 req/min per user, I'd recommend sliding window counter
       with Redis sorted sets. Want me to explain the algorithm?

You:   Yes, but keep it brief. I want to understand it, not study it.
```

```
AI:     Store each request as a member in a Redis sorted set, keyed by
        user ID, scored by timestamp. On each request:
        1. Remove entries older than 60 seconds
        2. Count remaining entries
        3. If count < 100, add the new request and allow
        4. If count >= 100, reject with 429

        The sorted set automatically handles the sliding window.

You:    Makes sense. But won't that use a lot of Redis memory for
        high-traffic users? 100 entries per user per minute, if we
        have 10K concurrent users that's a million sorted set members.

AI:     Good catch. For lower memory, use the sliding window counter
        approach instead — store just two counters per user (current
        window and previous window) and weight them:

        count = prev_window_count * overlap_ratio + current_window_count

        This uses 2 keys per user instead of N members. Want me to
        implement this version?

You:    Yes. Write the Express middleware. Include the rate limit
        headers (X-RateLimit-*) and make the limit configurable.

AI:     [generates the middleware]

You:    The implementation looks good but you're using separate Redis
        GET calls for the two window counters. Can we use a pipeline
        or MULTI to reduce round trips?

AI:     [refactors to use Redis pipeline, reducing from 4 round trips
        to 1]

You:    Perfect. One more thing — we need different limits for different
        API tiers (free: 60/min, pro: 300/min, enterprise: 1000/min).
        The user's tier is on req.user.tier.

AI:     [adds tier-based configuration lookup]
```

Notice the pattern. You set direction and constraints. AI proposed options with trade-offs. You caught a scalability issue AI missed. AI adapted. You spotted an optimization. AI implemented it. Neither of you could have produced the final result as efficiently alone.

The session took about 25 minutes. The rate limiter was production-ready, well-tested, and optimized — because the conversation surfaced issues a single prompt never would have.

When to Override AI

Pair programming means disagreeing sometimes. Here is when to override AI's suggestions:

When AI over-engineers. AI loves abstractions. If it suggests a strategy pattern for something that only has two variants, push back: "Keep it simple — a plain if/else is fine here."

When AI ignores your constraints. AI may suggest the ideal solution without considering your timeline, team skills, or existing architecture. "We don't have time for a full event-driven rewrite. What's the pragmatic 80% solution?"

When AI follows the wrong convention. AI may use popular patterns that don't match your team's conventions. "We don't use class-based services — use the functional pattern from our existing services."

When your instinct says no. Experienced developers carry intuition built from years of shipping software. If something feels wrong about AI's suggestion, trust your gut and investigate. More often than not, your instinct is pointing at a real problem.

When to Follow AI

Equally important — know when to defer to AI:

When AI knows the API better than you. If AI suggests a library function you haven't heard of, check if it exists. It often does, and it is often better than your hand-rolled version.

When AI catches an edge case you missed. "What about when the list is empty?" is exactly the kind of review comment you want from a pair partner.

When AI suggests a simpler approach. If your design is complex and AI offers something simpler that achieves the same goal, seriously consider the simpler version.

When AI identifies a pattern you should follow. "This is the same pattern as your existing user service — should we follow the same structure?" Yes, you should.

Building Pair Programming Habits

1. **Start every session with context.** Even if you think AI remembers, re-establish it.
2. **Think out loud.** Explain your reasoning. AI responds better to reasoning than to commands.
3. **Review every output.** Don't accept blindly. Read the code. Understand it.
4. **Commit frequently.** Small commits after each chunk. Easy to revert if needed.
5. **Take breaks from AI.** Sometimes you need to think without AI input. Close the tool and think.
6. **End sessions with a summary.** "Here's what we built today, here's what's left for tomorrow." This helps you (and AI) pick up where you left off.

Chapter 21: When AI Code Goes Wrong

Common Failure Modes, Hallucinated APIs, Subtle Bugs, Trust Boundaries

AI-generated code fails in specific, predictable ways. If you know the failure patterns, you can catch them before they ship. If you don't, they will find you in production — usually at the worst possible time.

This chapter is a field guide to AI code failures. Study it. Reference it. Build your review process around it.

Failure Mode 1: Hallucinated APIs

AI invents functions, methods, classes, and entire libraries that don't exist. This is not a rare edge case — it happens regularly, especially with less common libraries and newer APIs.

Examples

```
## AI-generated (WRONG) — this method doesn't exist
from fastapi import FastAPI
app = FastAPI()

@app.on_event("startup")
async def startup():
    # FastAPI deprecated on_event in favor of lifespan
    # but that's not the real problem...
    await app.state.db.initialize()  # .initialize() doesn't exist on state

// AI-generated (WRONG) — hallucinated React hook
import { useSearchParams, useRouteLoaderData } from 'react-router-dom';

const data = useRouteLoaderData('root');
// This function exists, but the usage pattern is wrong
// It requires specific loader configuration that AI didn't set up
```

```go
// AI-generated (WRONG) — mixed up library APIs
import "github.com/gin-gonic/gin"

func handler(c *gin.Context) {
    c.JSON(200, c.BindJSON(&request))
    // BindJSON returns an error, not the request
    // And you're sending the error as the response
}
```

How to Catch Them

1. **Verify every import.** If you don't recognize a class or function, check the library documentation.
2. **Watch for method chains that look too convenient.** If AI chains three methods together in a way that seems too elegant, verify each one exists.
3. **Be suspicious of newer library versions.** AI confidently uses APIs from the wrong version.
4. **Run the code.** Hallucinated APIs fail immediately at runtime. Don't skip the "actually run it" step.

Failure Mode 2: Subtly Wrong Logic

This is the most dangerous failure mode because the code runs, passes basic tests, and looks correct. The bug only surfaces under specific conditions.

The Off-by-One Family

```python
## AI-generated — spot the bug
def get_page(items: list, page: int, page_size: int) -> list:
    start = page * page_size
    end = start + page_size
    return items[start:end]

## Bug: If page is 1-indexed (as most APIs expect),
## page 1 returns items starting at index page_size, not 0
## Should be: start = (page - 1) * page_size
```

The Type Coercion Family

```javascript
// AI-generated — subtle type issue
function calculateDiscount(price, discountPercent) {
  return price * (1 - discountPercent / 100);
```

```
}

// Works fine in most cases. But:
calculateDiscount("49.99", 10)
// Returns NaN? No, worse: returns 44.991 (string coerced to number)
// But "49.99" * 0.9 = 44.991, which seems correct...
// Until you try: calculateDiscount("$49.99", 10) → NaN
```

The Edge Case Family

```
## AI-generated — works for most inputs
def find_median(numbers: list[float]) -> float:
    sorted_nums = sorted(numbers)
    mid = len(sorted_nums) // 2
    if len(sorted_nums) % 2 == 0:
        return (sorted_nums[mid - 1] + sorted_nums[mid]) / 2
    return sorted_nums[mid]

## Bug: What about empty list?
## sorted_nums = [], mid = 0, sorted_nums[0] → IndexError
## AI didn't add the empty check
```

The Time Zone Family

```
// AI-generated — looks correct
function isExpired(expirationDate: string): boolean {
  return new Date(expirationDate) < new Date();
}

// Bug: Date parsing without timezone specification
// "2026-03-15" is parsed as UTC midnight
// "new Date()" returns local time
// A user in UTC-8 at 11 PM on March 14 will see items
// expire 8 hours early
```

How to Catch Them

1. **Test edge cases explicitly.** Empty inputs, null values, zero, negative numbers, very large numbers, special characters.
2. **Test boundary conditions.** First element, last element, single element, maximum size.
3. **Think about types.** What if the input is a string instead of a number? What if it's undefined?

4. **Think about time.** Dates, timezones, daylight saving time, leap years, Unix epoch.
5. **Think about concurrency.** What if two requests hit this code at the same time?

Failure Mode 3: Confident Incorrectness

AI presents wrong information with the same confidence as correct information. It doesn't hedge. It doesn't say "I'm not sure." It states incorrect facts as if they were obvious.

```
## AI might confidently state:
## "Python's `dict.items()` returns a list of tuples"
##
## Wrong. In Python 3, it returns a view object, not a list.
## This matters if you're modifying the dict while iterating.
```

```
// AI might confidently generate:
// "Using Promise.allSettled to wait for all promises
//  and throw if any fail"
const results = await Promise.allSettled(promises);
if (results.some(r => r.status === 'rejected')) {
  throw new Error('Some operations failed');
}
// Problem: Promise.allSettled NEVER rejects. That's the point.
// AI confused it with Promise.all
```

How to Catch It

Trust but verify. When AI states something as fact, especially about language semantics, library behavior, or API contracts, verify against the official documentation. AI's confidence level has zero correlation with its correctness.

Failure Mode 4: Security Blind Spots

Covered in depth in Chapter 17, but worth repeating: AI-generated code has systematic security weaknesses.

The pattern: AI generates the functional path (make it work) and skips the security path (make it safe). Authentication, authorization, input validation, output encoding — AI includes these when asked, omits them when not asked.

```
// AI generates a perfectly functional file upload endpoint
// that accepts any file type, any size, stores it with the
// user-provided filename (path traversal), and serves it
// directly (XSS via uploaded HTML/SVG)

// All five of those issues need to be caught by a human reviewer
```

Failure Mode 5: Framework Misuse

AI knows many frameworks at a surface level. It generates code that uses the framework but doesn't follow its conventions, producing code that works but fights the framework.

```
## AI uses Django but fights it:
## Instead of Django's ORM relationships:
class OrderView(View):
    def get(self, request, order_id):
        order = Order.objects.get(id=order_id)
        items = OrderItem.objects.filter(order_id=order_id)  # Manual join
        customer = Customer.objects.get(id=order.customer_id)  # Manual join
        # Should be: order = Order.objects.select_related('customer')
        #                     .prefetch_related('items')
        #                     .get(id=order_id)

// AI uses React but fights it:
function UserList() {
  const [users, setUsers] = useState([]);

  // Anti-pattern: manual DOM manipulation in React
  useEffect(() => {
    const container = document.getElementById('user-list');
    users.forEach(user => {
      const div = document.createElement('div');
      div.textContent = user.name;
      container.appendChild(div);
    });
  }, [users]);

  return <div id="user-list" />;
  // Should just map over users in JSX
}
```

Failure Mode 6: Stale Patterns

AI's training data has a cutoff. Patterns that were correct in 2023 may be deprecated, insecure, or just outdated in 2026.

Common stale patterns AI generates: - React class components instead of functional components with hooks - Node.js `require()` instead of ES module `import` - Express 4 patterns in Express 5 projects - Python 2 compatibility patterns in Python 3.11+ projects - Outdated authentication patterns (sessions instead of JWTs, or vice versa, depending on context)

How to Catch It

Specify versions in your prompts: "Using React 18," "Node.js 20 with ESM," "Python 3.12." Check that generated code matches your actual dependency versions.

Failure Mode 7: The Copy-Paste Trap

AI sometimes generates code that works in isolation but fails in your project because it makes assumptions about your environment:

- Assumes a dependency exists that you don't have
- Assumes a global variable or configuration that your project doesn't define
- Generates TypeScript code with `strict: false` assumptions when your project uses `strict: true`
- Creates a function that duplicates existing functionality in your codebase

```
// AI generates a utility function for deep cloning
function deepClone<T>(obj: T): T {
  return JSON.parse(JSON.stringify(obj));
}
// Your project already has lodash.cloneDeep or structuredClone
// And JSON.parse/stringify doesn't handle dates, undefined,
// functions, or circular references
```

More Failure Examples with Actual Bad Code

The failure modes above are categories. Here are specific examples of bad code AI generates in each so you know exactly what to look for during review.

The Silent Data Loss Bug

```
// AI-generated function to update user preferences
async function updatePreferences(userId: string, newPrefs: Partial<Preferences>)
 {
  await prisma.user.update({
    where: { id: userId },
    data: { preferences: newPrefs },
  });
}

// Bug: If preferences is a JSON column and newPrefs is { theme: 'dark' },
// this REPLACES the entire preferences object.
// The user's existing preferences (language, timezone, notifications)
// are silently destroyed.
//
// Fix: merge with existing preferences first:
// data: { preferences: { ...existingUser.preferences, ...newPrefs } }
```

This bug is insidious because it passes every test where you only set preferences once. It only surfaces when a user updates one preference and loses the others. AI generates this pattern constantly because `update` with a partial object is the natural thing to write, and the ORM does exactly what you ask — it just doesn't do what you meant.

The Race Condition That Only Fails in Production

```
## AI-generated inventory check and purchase
async def purchase_item(user_id: str, product_id: str, quantity: int):
    product = await db.products.find_one(product_id)

    if product.stock < quantity:
        raise InsufficientStockError()

    # Time passes between the check and the update...

    await db.products.update_one(
        {"_id": product_id},
```

```
        {"$inc": {"stock": -quantity}}
    )
    await db.orders.insert_one({
        "user_id": user_id,
        "product_id": product_id,
        "quantity": quantity,
    })

## Bug: Two users can read stock=1 simultaneously, both pass the check,
## and both decrement stock. Final stock: -1. You just sold inventory
## you don't have.
##
## Fix: Use an atomic update with a condition:
## await db.products.update_one(
##     {"_id": product_id, "stock": {"$gte": quantity}},
##     {"$inc": {"stock": -quantity}}
## )
## Then check if the update matched any documents.
```

AI generates check-then-act patterns like this because they read clearly and work perfectly in sequential testing. The race only surfaces under concurrent load, which is exactly the condition AI never tests against.

The Encoding Bug That Corrupts Data

```
// AI-generated CSV export
function exportToCSV(rows) {
  const header = Object.keys(rows[0]).join(',');
  const body = rows.map(row =>
    Object.values(row).join(',')
  ).join('\n');
  return header + '\n' + body;
}

// Bug: If any value contains a comma, newline, or double quote,
// the CSV is corrupted. A company named "Smith, Jones & Co" becomes
// two columns instead of one.
//
// Fix: Properly escape values:
// const escape = (val) => `"${String(val).replace(/"/g, '""')}"`;
```

This is a "works in the demo" bug. AI tests with clean data. Your production data has commas in addresses, newlines in descriptions, and quotes in company names.

Building Trust Boundaries

A trust boundary is the line where you stop trusting AI output and start verifying.

Inside the trust boundary (high trust): - Boilerplate code (Express routes, React components, SQL migrations) - Well-known patterns with clear specifications - Code that will be immediately tested - Documentation and comments

At the trust boundary (verify carefully): - Business logic - Error handling - Data validation - Authentication and authorization

Outside the trust boundary (always verify): - Security-critical code - Financial calculations - Cryptographic operations - Data deletion operations - Production deployment scripts - Anything that, if wrong, causes irreversible damage

The more consequential the code, the harder you review it. AI-generated code for a signup form gets a quick review. AI-generated code for a payment processor gets a line-by-line audit.

The Failure Recovery Process

When AI-generated code fails in production:

1. **Fix the immediate issue.** This isn't the time for root cause analysis.
2. **Determine if AI-generated code was the cause.** Check commit history and AI assistance markers.
3. **Categorize the failure mode.** Which of the patterns above does it match?
4. **Add a test that catches this specific failure.** The test should have caught this before deployment.
5. **Update your review checklist.** Add the new failure pattern to your AI code review process.
6. **Update your AI configuration.** Add the rule to CLAUDE.md or .cursorrules so AI avoids the pattern next time.

This process turns each failure into a permanent improvement. Over time, your AI-assisted workflow grows more robust because each mistake sharpens the review process.

The Realist's Approach

AI code goes wrong. So does human code. The question isn't "does AI make mistakes?" but "does AI make mistakes at a rate and severity that outweighs the productivity benefit?"

The honest answer: yes, sometimes. For routine code, AI's error rate is low and the productivity benefit is enormous. For complex, security-critical, or domain-specific code, the error rate is higher and the verification effort is greater.

The key is calibrating your review effort to the risk level. Use AI freely for low-risk code. Use AI cautiously for high-risk code. And always, always run the tests.

One final thought: the developers who have the worst experience with AI-generated code are the ones who treat AI output as finished code. The developers who have the best experience treat AI output as a first draft — a starting point that gets them 80% of the way there, with the remaining 20% being the review, testing, and refinement that turns "works in the demo" into "works in production." That 20% is where your expertise earns its keep.

Defense to Offense

When AI code goes wrong is defense. Chapter 22 goes on offense — building AI-powered features into your own applications. APIs, SDKs, embeddings, and retrieval-augmented generation: how to add AI capabilities to the software you build.

Part VII: AI Products and Teams

"Software is a great combination between artistry and engineering."

— Bill Gates

Chapter 22: Building AI-Powered Features

Adding AI Capabilities to Your Own Applications (APIs, SDKs)

So far, this book has been about using AI to write code. This chapter is about writing code that uses AI. The distinction matters: you are no longer just consuming AI tools — you are building AI capabilities into the software your users interact with.

In 2026, AI-powered features are table stakes. Users expect smart search, natural language interfaces, content generation, and intelligent recommendations. This chapter covers how to add those features to your applications using AI APIs and SDKs.

The AI API Landscape

You have several options for adding AI capabilities:

OpenAI API: GPT-5.4, GPT-5.4 mini, embeddings, DALL-E for images. The largest ecosystem and broadest feature set.

Anthropic API (Claude): Claude Sonnet 4.6, Claude Haiku 4.5 for fast/cheap tasks. Strong on nuanced text analysis, code generation, and complex reasoning.

Google AI (Gemini): Gemini 3.1 Pro and Flash models. Excellent for multimodal tasks (text + image + video) and Google Cloud integration.

AWS Bedrock: Access to multiple models (Claude, Llama, Mistral) through one AWS service. Good for organizations already on AWS.

Local models (Ollama, vLLM): Run models on your own hardware. No data leaves your network. Lower quality than cloud models, but total privacy.

Your First AI Feature: Semantic Search

The most impactful AI feature for most applications is semantic search — finding results based on meaning, not just keywords.

Architecture

```
User query: "laptop that's good for video editing"

Traditional search: Searches for "laptop" AND "video" AND "editing"
→ Misses products described as "workstation" or "creative pro"

Semantic search: Converts query to a vector embedding,
finds products with similar embeddings
→ Finds relevant products regardless of exact keyword match
```

Implementation

```
import OpenAI from 'openai';
import { PgVector } from './lib/pgvector';

// Note: We use OpenAI for embeddings because Anthropic does not offer an
// embedding model. This is a common multi-provider pattern — use the best
// model for each task regardless of vendor. Chat completions might go
// through Anthropic (see the next section), while embeddings go through
// OpenAI or another provider that offers a dedicated embedding model.
const openai = new OpenAI();
const db = new PgVector();

// Step 1: Generate embeddings for your products (do once, at import)
async function embedProducts(products: Product[]): Promise<void> {
  for (const product of products) {
    const text = `${product.name} ${product.description} ${product.category}`;

    // Use a dedicated embedding model
    const embedding = await generateEmbedding(text);

    await db.query(
      'UPDATE products SET embedding = $1 WHERE id = $2',
      [JSON.stringify(embedding), product.id]
    );
  }
}
```

```
// Step 2: Search by semantic similarity
async function semanticSearch(
  query: string,
  limit: number = 10
): Promise<Product[]> {
  const queryEmbedding = await generateEmbedding(query);

  // Use pgvector's cosine similarity operator
  const results = await db.query(`
    SELECT *, 1 - (embedding <=> $1::vector) AS similarity
    FROM products
    WHERE 1 - (embedding <=> $1::vector) > 0.7  -- Minimum similarity threshold
    ORDER BY embedding <=> $1::vector
    LIMIT $2
  `, [JSON.stringify(queryEmbedding), limit]);

  return results.rows;
}

// Embedding generation (using OpenAI's embedding model)
async function generateEmbedding(text: string): Promise<number[]> {
  const response = await openai.embeddings.create({
    model: 'text-embedding-3-small',
    input: text,
  });

  return response.data[0].embedding;
}
```

Database Setup

```
-- Enable pgvector extension
CREATE EXTENSION IF NOT EXISTS vector;

-- Add embedding column to products
ALTER TABLE products ADD COLUMN embedding vector(1536);

-- Create index for fast similarity search
CREATE INDEX idx_products_embedding ON products
  USING ivfflat (embedding vector_cosine_ops)
  WITH (lists = 100);
```

Adding a Chat Interface

Basic Chat Completion

```
import Anthropic from '@anthropic-ai/sdk';

const anthropic = new Anthropic();

interface ChatMessage {
  role: 'user' | 'assistant';
  content: string;
}

async function chat(
  messages: ChatMessage[],
  systemPrompt: string
): Promise<string> {
  const response = await anthropic.messages.create({
    model: 'claude-sonnet-4-5',
    max_tokens: 1024,
    system: systemPrompt,
    messages: messages,
  });

  return response.content[0].type === 'text'
    ? response.content[0].text
    : '';
}

// Usage: Customer support chatbot
const systemPrompt = `You are a helpful customer support agent for
an e-commerce store. You can help with order status, returns,
and product questions.

Important rules:
- Never reveal internal system details
- For order issues, always ask for the order number
- Be friendly but concise
- If you don't know something, say so and offer to connect
  the user with a human agent`;

app.post('/api/chat', async (req, res) => {
  const { messages } = req.body;
```

```
  // Rate limiting and input validation here

  const response = await chat(messages, systemPrompt);

  res.json({ data: { message: response } });
});
```

Streaming Responses

For a better user experience, stream responses token by token:

```
app.post('/api/chat/stream', async (req, res) => {
  const { messages } = req.body;

  res.setHeader('Content-Type', 'text/event-stream');
  res.setHeader('Cache-Control', 'no-cache');
  res.setHeader('Connection', 'keep-alive');

  const stream = await anthropic.messages.stream({
    model: 'claude-sonnet-4-5',
    max_tokens: 1024,
    system: systemPrompt,
    messages: messages,
  });

  for await (const event of stream) {
    if (event.type === 'content_block_delta' &&
        event.delta.type === 'text_delta') {
      res.write(`data: ${JSON.stringify({ text: event.delta.text })}\n\n`);
    }
  }

  res.write('data: [DONE]\n\n');
  res.end();
});
```

Frontend Streaming Consumer

```
async function* streamChat(messages: ChatMessage[]): AsyncGenerator<string> {
  const response = await fetch('/api/chat/stream', {
    method: 'POST',
    headers: { 'Content-Type': 'application/json' },
    body: JSON.stringify({ messages }),
  });
```

```
  const reader = response.body!.getReader();
  const decoder = new TextDecoder();
  let buffer = '';

  while (true) {
    const { done, value } = await reader.read();
    if (done) break;

    buffer += decoder.decode(value, { stream: true });
    const lines = buffer.split('\n\n');
    buffer = lines.pop() || '';

    for (const line of lines) {
      if (line.startsWith('data: ')) {
        const data = line.slice(6);
        if (data === '[DONE]') return;
        const parsed = JSON.parse(data);
        yield parsed.text;
      }
    }
  }
}

// React component usage
function ChatMessage({ messages }: { messages: ChatMessage[] }) {
  const [response, setResponse] = useState('');

  useEffect(() => {
    let cancelled = false;

    async function run() {
      setResponse('');
      for await (const chunk of streamChat(messages)) {
        if (cancelled) break;
        setResponse(prev => prev + chunk);
      }
    }

    run();
    return () => { cancelled = true; };
  }, [messages]);

  return <div className="chat-response">{response}</div>;
}
```

Retrieval-Augmented Generation (RAG)

RAG combines your data with AI to generate accurate, contextual responses:

```
async function ragQuery(userQuestion: string): Promise<string> {
  // Step 1: Find relevant documents using semantic search
  const relevantDocs = await semanticSearch(userQuestion, 5);

  // Step 2: Build context from relevant documents
  const context = relevantDocs
    .map(doc => `[${doc.title}]: ${doc.content}`)
    .join('\n\n');

  // Step 3: Ask AI with the context
  const response = await anthropic.messages.create({
    model: 'claude-sonnet-4-5',
    max_tokens: 1024,
    system: `You are a helpful assistant. Answer questions based
      on the provided context. If the context doesn't contain
      enough information to answer, say so. Do not make up information.

      Always cite which document your answer comes from.`,
    messages: [
      {
        role: 'user',
        content: `Context:\n${context}\n\nQuestion: ${userQuestion}`,
      },
    ],
  });

  return response.content[0].type === 'text'
    ? response.content[0].text
    : '';
}
```

Content Generation

Generating Product Descriptions

```
async def generate_product_description(
    product_name: str,
    features: list[str],
    target_audience: str,
    tone: str = "professional"
```

```
) -> str:
    response = await anthropic.messages.create(
        model="claude-sonnet-4-5",
        max_tokens=500,
        messages=[{
            "role": "user",
            "content": f"""Write a product description for:

Product: {product_name}
Features: {', '.join(features)}
Target audience: {target_audience}
Tone: {tone}

Requirements:
- 2-3 paragraphs
- Highlight key benefits, not just features
- Include a call to action
- Do not make claims that aren't supported by the features listed"""
        }]
    )
    return response.content[0].text
```

Complete Example: Adding Semantic Search to a Product Catalog

Here is how to add semantic search to an existing e-commerce application, end to end. This is one of the most common AI features developers add, and it covers the full workflow from database setup to production deployment.

What you are building: A search box where users type "comfortable shoes for walking all day" and get relevant results, even if no product contains those exact words.

Step 1: Set up the vector database.

You already have PostgreSQL. Install the pgvector extension:

```
CREATE EXTENSION IF NOT EXISTS vector;

-- Add an embedding column to your existing products table
ALTER TABLE products ADD COLUMN embedding vector(1536);

-- Create an index for fast similarity search (do this AFTER populating embeddin
```

```
gs)
CREATE INDEX idx_products_embedding
  ON products USING hnsw (embedding vector_cosine_ops);
```

Step 2: Build the embedding pipeline.

```
import OpenAI from 'openai';
import { Pool } from 'pg';

const openai = new OpenAI();
const db = new Pool();

// Generate embedding text from product data
function productToEmbeddingText(product: Product): string {
  // Combine fields that describe the product semantically
  return [
    product.name,
    product.description,
    product.category,
    product.tags?.join(', '),
    product.brand,
  ].filter(Boolean).join(' | ');
}

// Embed a single product
async function embedProduct(product: Product): Promise<number[]> {
  const text = productToEmbeddingText(product);
  const response = await openai.embeddings.create({
    model: 'text-embedding-3-small',
    input: text,
  });
  return response.data[0].embedding;
}

// Batch embed all products (run once, then on new product creation)
async function embedAllProducts(): Promise<void> {
  const { rows: products } = await db.query(
    'SELECT * FROM products WHERE embedding IS NULL'
  );

  // Process in batches of 100 (OpenAI embedding API supports batch input)
  for (let i = 0; i < products.length; i += 100) {
    const batch = products.slice(i, i + 100);
    const texts = batch.map(productToEmbeddingText);
```

```
    const response = await openai.embeddings.create({
      model: 'text-embedding-3-small',
      input: texts,
    });

    // Update each product with its embedding
    for (let j = 0; j < batch.length; j++) {
      await db.query(
        'UPDATE products SET embedding = $1 WHERE id = $2',
        [JSON.stringify(response.data[j].embedding), batch[j].id]
      );
    }

    console.log(`Embedded ${Math.min(i + 100, products.length)} / ${products.len
gth}`);
  }
}
```

Step 3: Build the search endpoint.

```
async function searchProducts(query: string, limit = 20): Promise<SearchResult[]>
 {
  // Generate embedding for the search query
  const response = await openai.embeddings.create({
    model: 'text-embedding-3-small',
    input: query,
  });
  const queryEmbedding = response.data[0].embedding;

  // Find similar products using cosine distance
  const { rows } = await db.query(`
    SELECT id, name, description, price, category,
           1 - (embedding <=> $1::vector) AS similarity
    FROM products
    WHERE embedding IS NOT NULL
    ORDER BY embedding <=> $1::vector
    LIMIT $2
  `, [JSON.stringify(queryEmbedding), limit]);

  return rows;
}

// Combine with traditional search for best results
async function hybridSearch(query: string, limit = 20): Promise<SearchResult[]>
{
```

```
  const [semanticResults, keywordResults] = await Promise.all([
    searchProducts(query, limit),
    db.query(`
      SELECT id, name, description, price, category,
             ts_rank(search_vector, plainto_tsquery($1)) AS rank
      FROM products
      WHERE search_vector @@ plainto_tsquery($1)
      ORDER BY rank DESC LIMIT $2
    `, [query, limit]),
  ]);

  // Merge and deduplicate, boosting products that appear in both
  const merged = mergeResults(semanticResults, keywordResults.rows);
  return merged.slice(0, limit);
}
```

Step 4: Hook the new product pipeline.

Add embedding generation to your product creation flow so new products are searchable immediately:

```
// In your product creation service
async function createProduct(data: CreateProductInput): Promise<Product> {
  const product = await prisma.product.create({ data });

  // Generate embedding asynchronously (don't block the response)
  queue.add('embed-product', { productId: product.id });

  return product;
}
```

The total implementation is about 150 lines of code. The quality jump over keyword search is dramatic — users who search "gift for someone who likes cooking" will find kitchen gadgets, cookbooks, and aprons, none of which contain the word "gift."

What to watch for: Embedding API calls add latency to search (typically 100-200ms). Cache embeddings for common queries in Redis. Also, the `text-embedding-3-small` model costs about $0.02 per million tokens — cheap, but track your usage.

Cost Management

AI API calls cost money. Manage costs carefully:

```
// Cost-aware model selection
function selectModel(task: string): string {
  switch (task) {
    case 'classification':
    case 'extraction':
    case 'simple_qa':
      return 'claude-haiku-4-5-20251001';   // Fast, cheap

    case 'generation':
    case 'analysis':
      return 'claude-sonnet-4-5';  // Balanced

    case 'complex_reasoning':
    case 'code_generation':
      return 'claude-opus-4-7';    // Expensive, best quality

    default:
      return 'claude-sonnet-4-5';
  }
}

// Implement caching
import { createHash } from 'crypto';

const responseCache = new Map<string, { response: string; expiry: number }>();

async function cachedAICall(
  prompt: string,
  model: string,
  cacheTTL: number = 3600000  // 1 hour default
): Promise<string> {
  const cacheKey = createHash('sha256')
    .update(`${model}:${prompt}`)
    .digest('hex');

  const cached = responseCache.get(cacheKey);
  if (cached && cached.expiry > Date.now()) {
    return cached.response;
  }
```

```
  const response = await callAI(prompt, model);
  responseCache.set(cacheKey, {
    response,
    expiry: Date.now() + cacheTTL
  });

  return response;
}
```

Error Handling for AI APIs

```
async function robustAICall(
  messages: ChatMessage[],
  options: { maxRetries?: number; timeout?: number } = {}
): Promise<string> {
  const { maxRetries = 3, timeout = 30000 } = options;

  for (let attempt = 0; attempt < maxRetries; attempt++) {
    try {
      const controller = new AbortController();
      const timeoutId = setTimeout(() => controller.abort(), timeout);

      const response = await anthropic.messages.create({
        model: 'claude-sonnet-4-5',
        max_tokens: 1024,
        messages,
      });

      clearTimeout(timeoutId);
      return response.content[0].type === 'text'
        ? response.content[0].text : '';

    } catch (error) {
      if (error instanceof Anthropic.RateLimitError) {
        // Back off exponentially
        const delay = Math.pow(2, attempt) * 1000;
        await new Promise(resolve => setTimeout(resolve, delay));
        continue;
      }

      if (error instanceof Anthropic.APIError && error.status >= 500) {
        // Server error, retry
        continue;
      }
```

```
      // Non-retryable error
      throw error;
    }
  }

  throw new Error('AI service unavailable after maximum retries');
}
```

Safety and Guardrails

When exposing AI to end users, you need guardrails:

```
// Content filtering
async function safeAIResponse(
  userInput: string
): Promise<{ response: string; flagged: boolean }> {
  // Step 1: Check user input for policy violations
  const inputCheck = await anthropic.messages.create({
    model: 'claude-haiku-4-5-20251001',
    max_tokens: 10,
    messages: [{
      role: 'user',
      content: `Is this message appropriate for a customer support chatbot?
        Answer only "yes" or "no". Message: "${userInput}"`
    }]
  });

  if (inputCheck.content[0].type === 'text' &&
      inputCheck.content[0].text.toLowerCase().includes('no')) {
    return {
      response: "I'm sorry, I can only help with product and order questions.",
      flagged: true,
    };
  }

  // Step 2: Generate response with safety system prompt
  const response = await chat([{ role: 'user', content: userInput }],
    safetySystemPrompt);

  return { response, flagged: false };
}
```

What AI Gets Wrong with AI Integration

Ignoring latency: AI API calls take 1–10 seconds. AI-generated code often calls APIs synchronously in request handlers, causing timeouts. Use background jobs, streaming, or caching.

Ignoring cost: A feature that calls GPT-4 on every page load will bankrupt you. Calculate expected costs before deploying.

Trusting AI output in production: AI responses can contain harmful, incorrect, or inappropriate content. Filter, validate, and constrain AI output before showing it to users.

No fallback: When the AI API is down (and it will be), what does your feature do? Always have a graceful degradation path.

Building AI features is one developer's job. Getting an entire team to use AI well is a leadership challenge. Chapter 23 covers team AI adoption — how to introduce AI tools to a development team, establish best practices, and measure the impact.

Chapter 23: Team AI Adoption

Introducing AI Tools to Your Development Team

You have been using AI coding tools for a while now. You are faster. Your code reviews are sharper. You reach for Claude Code or GitHub Copilot the way you reach for your IDE — it is just part of how you work. Now someone (your manager, your CTO, or maybe the voice in your head) is asking: "How do we get the whole team doing this?"

This is a different problem than learning AI tools yourself. Individual adoption is about skill. Team adoption is about culture, process, trust, and a surprising amount of organizational psychology. Most team AI rollouts fail not because the tools are bad but because the introduction was botched.

This chapter is the playbook for doing it right.

Why Team Adoption Is Harder Than Individual Adoption

When you adopted AI tools, you controlled every variable. You chose which tool to try, which tasks to use it on, how much to trust the output. You could experiment quietly and only show people the wins.

A team rollout is different for three reasons:

Skill variance is enormous. Some developers take to AI immediately. Some are skeptical but open. Some feel threatened. All three groups need different things. Treating them the same guarantees you alienate at least one.

Process changes affect everyone. If AI-generated code enters your codebase, every reviewer needs to know how to evaluate it. Your testing strategy may need to change. Your code review checklist definitely does. These are team-wide process shifts, not individual preferences.

Trust requires verification infrastructure. "I used AI for this" is useless as a code review comment if the reviewer has no framework for evaluating AI-generated code differently from human-written code. You need shared standards for when AI is appropriate, what verification looks like, and how to document AI assistance.

The Four Phases of Team Adoption

Phase 1: Controlled Pilot (Weeks 1–4)

Don't roll AI tools out to the entire team at once. Start with a pilot group of 3–5 developers who represent different experience levels and attitudes toward AI.

Selecting your pilot group:

- Include at least one skeptic. Skeptics find real problems enthusiasts overlook.
- Include at least one senior developer. Their buy-in influences the rest of the team.
- Include at least one junior developer. They use the tools differently and surface different issues.
- Exclude anyone actively hostile. Mandating participation breeds resentment. Let the hostile group come around on their own when they see the results.

What the pilot group does:

- Use AI tools on real work, not toy projects. The point is to learn how AI fits into your actual workflow.
- Keep a shared log of wins, failures, and surprises. A Slack channel or shared doc works.
- Meet weekly to compare notes. What worked? What produced garbage? What was unexpectedly useful?
- Track time savings honestly. Not "I feel faster" but "this task that usually takes two hours took 45 minutes."

What you are looking for:

- Which tools work best for your stack and workflow
- Where AI creates the most value (it is almost never where you predict)
- What guardrails you need (more below)

- What training the broader team will need

Phase 2: Standards and Guardrails (Weeks 3–6)

Based on pilot findings, establish team standards before expanding. This is the step most teams skip, and it is why most rollouts devolve into chaos.

Create an AI usage policy. This doesn't need to be a 50-page document. A single page covering:

- **Approved tools:** Which AI tools are sanctioned. This matters for security (data sent to AI providers) and for consistency (everyone on the same tools builds shared knowledge).
- **Approved use cases:** Code generation, test writing, documentation, debugging. Be explicit about what is encouraged.
- **Restricted use cases:** Security-sensitive code, code that handles PII, code that implements business-critical logic without human review. Be explicit about what needs extra scrutiny.
- **Prohibited use cases:** Pasting proprietary algorithms into public AI tools, using AI to generate code for compliance-regulated systems without review, copy-pasting AI output into production without any verification.

Example AI usage policy:

```
## AI Coding Tools — Team Policy

### Approved Tools
- Claude Code (via Anthropic API, business plan)
- GitHub Copilot (organization license)
- ChatGPT Plus (individual licenses, do NOT paste proprietary code)

### Encouraged Uses
- Boilerplate generation (CRUD endpoints, component scaffolding)
- Test generation (unit tests, integration test skeletons)
- Documentation (docstrings, README updates, API docs)
- Debugging assistance (error analysis, log interpretation)
- Code review preparation (pre-review self-check)
- Refactoring (extract method, rename, restructure)

### Requires Senior Review
- Database migration generation
- Security-related code (auth, encryption, access control)
```

```
- Infrastructure changes (Terraform, Kubernetes configs)
- Any AI-generated code touching payment processing

### Prohibited
- Pasting proprietary business logic into ChatGPT free tier
- Using AI output in production without code review
- AI-generated code in compliance-regulated modules without
  compliance team sign-off
```

Update your code review checklist. Add AI-specific items:

```
### Code Review Checklist — AI-Assisted Code

- [ ] Does the AI-generated code handle edge cases for our domain?
- [ ] Are there any hallucinated APIs, methods, or packages?
- [ ] Does the code follow our naming conventions and patterns?
- [ ] Has the developer verified the code works, not just compiles?
- [ ] Are tests included (not just the happy path)?
- [ ] Is the AI-generated code consistent with surrounding code style?
- [ ] For security-sensitive code: has a security-aware reviewer approved?
```

Phase 3: Team Rollout (Weeks 5–8)

Now expand to the full team. The pilot members become your internal champions — they have real experience and can help others through the learning curve.

Run hands-on workshops, not presentations. Nobody learns AI tools from a slide deck. Run 90-minute sessions where developers:

1. Set up the tools on their machines (have pilot members available to troubleshoot)
2. Work through three exercises on their own code:
    - Generate a function they were planning to write anyway
    - Write tests for existing untested code
    - Debug a real bug using AI assistance
3. Share what they learned with the group

Pair program with AI. For the first two weeks, encourage developers to pair with a pilot member on AI-assisted tasks. Watching someone use AI effectively teaches more than any documentation.

Set realistic expectations. Tell the team explicitly:

- The first week will be slower, not faster. Every new tool has a productivity dip.
- AI will produce wrong code sometimes. That is normal, not a failure of the tool.
- Finding the right prompting style takes practice. Share prompt patterns that worked during the pilot (see Appendix B).
- It is fine to skip AI on some tasks. Some things are faster to write yourself.

Phase 4: Measurement and Iteration (Ongoing)

You can't improve what you don't measure. Track these metrics:

Quantitative metrics:

- **Cycle time:** Time from first commit to PR merge. This should drop.
- **PR throughput:** PRs merged per developer per week. This should rise.
- **Bug rate:** Bugs per release. This should not increase (if it does, your review process needs work).
- **Test coverage:** Coverage percentage. This should rise (AI makes writing tests much easier).
- **Time to first commit:** How long new team members take to ship their first meaningful contribution. AI dramatically accelerates onboarding.

Qualitative metrics:

- Developer satisfaction surveys (monthly, keep them short)
- Code review feedback (are reviewers flagging more AI-specific issues?)
- Sprint retrospective themes (is AI coming up positively or negatively?)

What to watch for:

- **Over-reliance.** If developers stop understanding the code they ship, you have a problem. Spot-check by asking developers to explain their PRs without looking at the code.
- **Cargo culting.** Developers copying AI patterns without understanding why. This shows up as code that works but is architecturally wrong for your system.

- **Review fatigue.** If AI bumps PR volume by 50%, reviewers drown. You may need to adjust review processes or rotate review duties.

Handling Resistance

You will encounter resistance. Here is how to handle the three most common types.

"AI Is Going to Replace Us"

This fear is real and dismissing it is counterproductive. Address it directly:

"AI tools make developers more productive, which makes developers more valuable, not less. Companies adopting AI are shipping more software, not hiring fewer developers. The developers at risk aren't the ones using AI — they are the ones whose competitors are using AI while they are not."

Back this up with data. Every major study on AI coding tools shows that organizations expand their development ambitions when developers get faster. More features get built. More experiments get tried. The bottleneck shifts from writing code to deciding what to build.

"The Code Quality Will Drop"

This is the concern of your best engineers, and they aren't wrong to raise it. AI-generated code can be subtly wrong in ways that are hard to catch. The answer is not "trust the AI" — it is "trust the process."

Show them the guardrails you built in Phase 2. Demonstrate that AI-generated code goes through the same review process (or a more rigorous one) as human-written code. Invite them to be the senior reviewers who provide that extra scrutiny. Channel their skepticism into constructive oversight.

"I Don't Need It"

Some developers are genuinely fast and productive without AI. Don't force adoption. Instead:

- Make the tools available and supported.
- Share team metrics showing the impact.

- Let social proof do its work. When they see peers shipping faster at equal quality, most holdouts come around naturally.
- Respect that some developers will never adopt AI tools, and that's fine as long as their output remains competitive.

Configuring Tools for Teams

Individual AI tool setups don't scale. Teams need shared configurations.

Shared Prompt Libraries

Create a repository of prompts that work for your codebase:

```
## Team Prompt Library

### New API Endpoint
"Create a [METHOD] endpoint at [PATH] that [DESCRIPTION].
Follow our standard patterns in src/routes/ – use the
validation middleware, return standard ApiResponse format,
include error handling for [SPECIFIC CASES]. Write the
route handler, service function, and unit tests."

### Database Migration
"Write a migration that [DESCRIPTION]. Use our migration
format (see migrations/README.md). Include both up and
down migrations. Consider the impact on existing data –
if there are rows that would violate new constraints,
include a data migration step."

### Code Review Prep
"Review this code for: security issues, performance
problems, missing error handling, deviation from our
coding standards (see CONTRIBUTING.md), and missing
test cases. Format your review as a list of findings
with severity (critical/major/minor/nit)."
```

Shared Configuration Files

For tools that support project-level configuration:

```
## CLAUDE.md (in your repo root)

### Project Context
```

```
This is a Node.js/TypeScript monorepo using:
- Express for API routes
- Prisma for database access
- Jest for testing
- Our custom middleware stack (see src/middleware/)

### Conventions
- All API responses use ApiResponse<T> wrapper
- Database queries go through service layer, never in routes
- Tests live next to the code they test (foo.ts / foo.test.ts)
- Error handling uses our AppError class hierarchy

### Do Not
- Generate raw SQL — always use Prisma
- Skip input validation on any endpoint
- Use any/unknown types without justification
- Import from internal packages using relative paths
```

License Management

Ensure all developers are on appropriate license tiers:

Tool	Recommended Tier	Why
Claude Code	Max (team)	Full context window, extended thinking, team management
GitHub Copilot	Business	Indemnification, admin controls, audit logs
Cursor	Business	Team settings sync, centralized billing

The business tiers aren't just about features — they provide admin controls, usage auditing, and IP indemnification that individual tiers don't. For any organization with more than a handful of developers, these are non-negotiable.

Measuring ROI

Your leadership will want numbers. Here is how to build the case.

Before/after cycle time: Measure average PR cycle time for four weeks before rollout and four weeks after. Typical improvement: 20–40%.

Test coverage delta: Measure coverage before and after. AI makes generating tests so easy that coverage typically jumps 10–20 percentage points.

Developer satisfaction: Run a simple survey. "How satisfied are you with your development tooling?" on a 1–5 scale. Measure before and after.

Cost calculation:

Not every developer needs every tool. A common pattern: give everyone GitHub Copilot Business ($19/month) for inline autocomplete and give your power users — senior developers, team leads, and architects doing complex multi-file work — Claude Code Max ($100/month) on top. Here is what that looks like for a team of 20 where half get both tools:

```
Copilot Business for all 20 developers:  20 x $19  = $380/month
Claude Code Max for 10 power users:      10 x $100 = $1,000/month
Monthly tool spend:                                   $1,380/month

Average developer salary: $150,000/year = $12,500/month
20% productivity improvement = $2,500/month effective value per developer
20 developers = $50,000/month effective value

ROI: $50,000 / $1,380 ≈ 36x return
```

The math is so favorable that cost is almost never the real objection. Even if you give every developer both tools ($119/month per developer, $2,380/month for 20), the ROI is still over 20x. The real objections are cultural, and this chapter has given you the tools to address them.

Common Mistakes in Team Rollouts

Mandating specific tools without team input. Let the team have a voice in tool selection. Run the pilot with 2–3 options and let usage data drive the decision.

Skipping the pilot phase. "Let's just buy everyone licenses and see what happens" produces uneven adoption, inconsistent practices, and frustration. The pilot phase exists to surface problems while they are small.

No guardrails. AI tools without guardrails are like giving everyone commit access to main. The power is real, but so is the risk. Build the guardrails first.

Measuring the wrong things. Lines of code per day is not a useful metric. Neither is "number of AI prompts." Measure outcomes: cycle time, bug rate, test coverage, developer satisfaction.

Forgetting about onboarding. Once AI tools are part of your workflow, new hires need to learn them too. Fold AI tool training into your onboarding process. Pair new developers with experienced AI users during their first sprint.

Part VIII: Reference

"It always takes longer than you expect, even when you take into account Hofstadter's Law."

— Douglas Hofstadter

Chapter 24: The Future of AI-Assisted Development

What's Coming, What Skills Matter, What Won't Change

Prediction is a fool's game, especially in AI. Six months ago, the capabilities you use today would have seemed unlikely. Six months from now, this chapter might read like someone in 2005 predicting that phones would one day have cameras.

But there are patterns. There are trajectories. And there are truths about software engineering that no amount of AI progress will change. This chapter separates signal from noise.

Where the Tools Are Heading

Longer Context, Deeper Understanding

The biggest near-term improvement is not smarter models — it is bigger context windows and better use of them. In early 2024, a 100K token context window was impressive. By the time you read this, million-token contexts are standard, and the tools are getting better at using all of that space.

What this means for you: AI tools will understand your entire codebase, not just the file you are editing. You will say "refactor the authentication system to use OAuth2" and the tool will understand every file that touches authentication, every test that covers it, and every downstream service that depends on it. The era of handing AI a single function and hoping for the best is ending.

Agentic Development

The biggest shift underway is the move from AI as a tool you query to AI as an agent that executes. Instead of asking "write me a function that does X" and getting back text, you say "implement feature X" and the AI will:

1. Read the relevant code
2. Plan the changes across multiple files

3. Write the code
4. Run the tests
5. Fix what fails
6. Open a pull request

This is not hypothetical. Claude Code already works this way for many tasks. GitHub Copilot Workspace, Cursor's agent mode, and similar tools are converging on the same pattern. The question is not whether agentic development becomes standard but how fast reliability reaches the point where you trust it for non-trivial work.

What to watch for: The critical metric is not "can the agent complete the task" but "how often does the agent complete the task correctly on the first try." When that number crosses 80% for your typical tasks, agentic workflows become the default. Until then, they are a productivity tool that still needs supervision.

Multi-File, Multi-Step Reasoning

Current tools already understand relationships across files. Future tools will be dramatically better at multi-step reasoning — understanding that changing a database schema requires updating the ORM model, the API endpoint, the frontend form, the validation logic, the tests for all of the above, and the migration script.

This is the difference between AI that writes code and AI that does engineering. We are in the transition between those worlds right now.

Real-Time Collaboration

Expect AI to move from a personal assistant model to a team collaboration model. Instead of each developer carrying their own AI context, teams will share AI contexts that understand the entire project — its history, its conventions, and its current state.

Imagine an AI that joins your standup, knows what everyone is working on, flags merge conflicts before they happen, and suggests task assignments based on expertise and current workload. The technology to build this exists today. The products that package it well are coming.

Specialized Models

The era of "one model does everything" is giving way to specialized models for specific tasks:

- **Code review models** trained specifically on code review patterns, security vulnerabilities, and best practices
- **Test generation models** that understand coverage gaps and edge case patterns
- **Migration models** that specialize in upgrading frameworks, languages, and dependencies
- **Documentation models** that maintain consistency across large documentation sets

You will not choose a single AI provider. You will use an ecosystem of specialized models, each best at its task, orchestrated by your IDE or development platform.

What Skills Will Matter More

Architecture and System Design

As AI handles more of the code generation, the premium on knowing what to build rises. Architecture — understanding trade-offs, designing for scale, choosing the right patterns for the right problems — becomes the highest-value skill a developer can have.

AI can generate a microservice in minutes. Knowing whether you should use a microservice or a monolith, how to design service boundaries, how to handle distributed transactions, how to plan for failure — that is human work, and it is the work that matters most.

Invest in: System design practice, distributed systems, studying real-world architecture decisions (and their consequences).

Specification and Requirements

The better AI gets at turning specifications into code, the more valuable the skill of writing precise specifications becomes. A vague spec produces vague code. A precise spec produces correct code.

This sounds obvious, but most developers have never had to write truly precise specifications because they were the ones implementing them. When you are the person writing the code, ambiguity resolves itself — you make a decision and move on. When AI is writing the code, ambiguity produces bugs.

Invest in: Technical writing, decomposing complex requirements into unambiguous specifications, practicing prompt engineering as a form of specification writing.

Verification and Testing

If code generation accelerates by 5x but verification stays the same, verification becomes the bottleneck. Developers who can efficiently verify AI-generated code — through testing, code review, and systematic validation — will be more valuable than ever.

This means going beyond "does it compile and pass existing tests." It means understanding what the code should do, what edge cases exist, what security implications the changes have, and what failure modes are possible. Testing is no longer just QA work — it is a core engineering discipline.

Invest in: Property-based testing, mutation testing, security testing, reading code critically (not just for correctness but for subtle design flaws).

Domain Expertise

AI knows everything in general and nothing in particular. It has read every Stack Overflow answer about payment processing but has never processed an actual payment. It knows the theory of HIPAA compliance but does not understand your specific compliance requirements.

Domain expertise — deep knowledge of your industry, your users, your regulatory environment, and your business — becomes the differentiator that separates a developer from a prompt jockey. The developer who understands healthcare data workflows will always ship better healthcare software than the developer who just types better prompts.

Invest in: Understanding your industry deeply. Talk to users. Read regulations. Learn the business, not just the code.

AI Fluency

A meta-skill is emerging: the ability to work with AI effectively. This is not just prompt engineering (though that is part of it). It is knowing:

- When to use AI and when to code manually
- How to decompose problems into AI-appropriate chunks
- How to verify AI output efficiently
- How to provide context that maximizes output quality
- How to recognize when AI is confidently wrong
- How to combine multiple AI tools for different parts of a workflow

CS programs do not teach this skill yet. You are learning it by reading this book and by practice. Within a few years, AI fluency will be as expected on a developer resume as Git proficiency.

What Skills Will Matter Less

Memorization of Syntax and APIs

Knowing the exact signature of `Array.prototype.reduce()` or remembering which HTTP status code means "Gone" — these were never high-value skills, and AI has made them essentially worthless. Describe what you want, and the AI will get the syntax right.

This does not mean you should skip the concepts. Knowing what a reduce operation does, when to use it, and what its performance characteristics are — that still matters. Knowing the exact syntax does not.

Boilerplate Fluency

The ability to bang out Express route handlers, React component scaffolding, Terraform configurations, and CI pipeline YAML from memory is no longer a competitive advantage. AI generates this code instantly and correctly. Time spent memorizing boilerplate is time better spent elsewhere.

Single-Language Depth

Being "a Python developer" or "a Java developer" made sense when switching languages was expensive. AI has made language boundaries nearly irrelevant. A competent developer with AI assistance can ship production-quality code in any mainstream language. The value shifts from knowing one language deeply to understanding programming concepts that transcend any single language.

What Will Never Change

Code Is a Liability

Every line of code is a line that can break, a line that needs to be maintained, and a line that someone has to understand. The best code is no code. The second best is the minimum code that solves the problem.

AI makes it trivially easy to generate large amounts of code. That makes it more tempting than ever to over-engineer, to add features that are not needed, to build abstractions before they are justified. The discipline of writing less code — of asking "do we actually need this?" — becomes more important, not less, when generating code is cheap.

Reading Code Is Harder Than Writing Code

This has been true since the first line of code was written, and it remains true in the AI era. AI can generate code in seconds that takes minutes to review. The asymmetry between generation and comprehension means code review, code reading, and code understanding remain the bottleneck skills.

If anything, AI makes this worse. A developer who does not use AI understands their code because they wrote it. A developer who uses AI to generate code might not fully understand what was generated. The discipline of reading every line, understanding every line, and being able to explain every line is non-negotiable.

Fundamentals Win

Data structures and algorithms. Networking. Operating systems. Database internals. Security principles. Distributed systems. Concurrency. These fundamentals have been relevant for 50 years, and they will be relevant for 50 more.

AI does not change what makes software work. It changes how fast you can write it. A developer who understands why a hash map is O(1) for lookups and when it degrades will make better decisions than a developer who lets AI choose data structures without grasping the trade-offs.

The developers who thrive in the AI era will be the ones with strong fundamentals and AI fluency — not the ones who traded fundamentals for prompt tricks.

Software Is a Human Activity

Software exists to serve people. Understanding what people need, translating that into systems, and iterating based on feedback — this is fundamentally human work. AI can help at every step, but the empathy, judgment, and creativity at the core of great software engineering come from you.

The best developers have always been the ones who talk to users, understand the problem before reaching for a solution, and push back when the proposed solution is wrong. AI does not change that. It amplifies it.

The Jobs Question

Let's address it directly, because it is on everyone's mind.

AI will not eliminate software development jobs in the foreseeable future. It will reshape them. Here is the honest assessment:

Jobs that will grow: Architecture and system design roles. AI/ML integration specialists. Security engineers (more code means more attack surface). Quality engineers and test architects. Developer-experience engineers who build AI-augmented workflows.

Jobs that will shrink: Roles that mostly consist of translating well-defined specifications into code. If your job is "take this Jira ticket and write the CRUD endpoint," AI is coming for it — not by replacing you, but by making you fast enough to take on higher-level work.

Jobs that will transform: Most software development roles. You will write less code directly and spend more time on design, verification, integration, and communication. The title stays the same. The job description shifts significantly.

The historical analogy is not "AI replaces programmers" — it is "high-level languages replaced assembly programmers." Assembly programmers did not disappear. They became C programmers. C programmers became Python programmers. Each transition raised the abstraction level and made the remaining work more about design and less about mechanics. AI is the next step in that progression.

Preparing for the Next Two Years

Here is your practical action plan:

1. **Master one agentic AI tool.** Pick Claude Code, Cursor, or Windsurf and learn it deeply. Agentic workflows will be the default within 18 months. Get ahead of the curve.

2. **Build your verification skills.** Learn property-based testing. Practice code review. Get comfortable with security scanning tools. As generation speeds up, verification is where you add value.

3. **Invest in architecture.** Read "Designing Data-Intensive Applications." Study system design interviews — not for the interviews but for the thinking patterns. Take on architecture work at your job.

4. **Go deeper in your domain.** Whatever industry you work in, become an expert in it. Domain knowledge is the moat AI cannot cross.

5. **Stay current, but not anxious.** New AI tools and capabilities will launch constantly. You do not need to try every one. Pick your tools, learn them well, and upgrade when the improvement justifies the switching cost.

6. **Teach others.** The best way to solidify your own understanding is to help your team adopt AI effectively. Chapter 23 gave you the playbook. Use it.

Final Thought

The developers who thrive in the AI era will not be the ones who use AI the most or the ones who resist it the longest. They will be the ones who understand what AI is good at, what they are good at, and how to combine both effectively.

You have spent 24 chapters learning exactly that. The epilogue wraps up with parting thoughts.

Epilogue: The Developer You're Becoming

You started this book with a question most developers are quietly asking: "How do I use AI without losing what makes me good at my job?"

Here is the answer, after 24 chapters of practical techniques, code examples, workflows, and hard-won lessons: you don't lose what makes you good. You amplify it.

The developer who understands system design uses AI to implement those designs faster. The developer who writes thorough tests uses AI to generate even more thorough tests. The developer who reviews code carefully uses AI to catch what human eyes miss. The developer who cares about clean architecture uses AI to refactor without fear.

AI does not replace your skills. It gives them leverage.

What You Have Learned

This book covered a lot of ground. You now know how to:

- Choose the right AI tools for your workflow and configure them for maximum effectiveness
- Generate code that actually works — not just compiles, but handles edge cases and follows your project's conventions
- Use AI for code review, debugging, testing, and documentation without over-relying on it
- Apply AI across domains — frontend, backend, DevOps, databases, mobile, API design
- Work with AI as a pair programmer, not just a code generator
- Recognize when AI is wrong, confidently wrong, and how to catch it
- Build AI-powered features into your own applications
- Lead your team through AI adoption without the chaos

That is a substantial toolkit. But the tools will change. New models will launch. New products will emerge. What will not change is the approach: understand what AI is good at, apply it to the right problems, verify the output, and keep thinking.

The Compound Effect

Productivity gains from AI are not linear — they compound. A developer who writes code 30% faster also reviews faster, iterates faster, and ships faster. Faster shipping means faster feedback. Faster feedback means better products. Better products mean more ambitious projects. More ambitious projects mean more interesting work.

The developers who adopted AI early — and adopted it well — are not just faster than they were before. They are working on different kinds of problems. Problems they never had time for when they were writing boilerplate by hand.

You are now one of those developers.

A Word of Caution

With great leverage comes the temptation to move too fast. AI makes it easy to generate enormous amounts of code in very little time. Resist the urge to equate volume with progress. Every line of code, whether written by you or generated by AI, is a line that must be understood, tested, maintained, and eventually replaced.

The discipline of saying "we don't need this yet" matters more than ever. The discipline of reading every line before committing it matters more than ever. The discipline of testing not just the happy path but the ugly edge cases matters more than ever.

Speed without discipline is just faster failure.

What Comes After This Book

Practice. That is it. That is the answer.

Read the appendices — the prompt patterns in Appendix B will accelerate your daily work, and the tool comparison in Appendix A will help you make informed decisions as the landscape evolves. But the real learning happens when you sit down with your codebase, your AI tool, and a real problem to solve.

Start small. Refactor a function. Generate tests for untested code. Write the documentation you have been putting off. Then move to something bigger. Implement a feature end-to-end with AI assistance. Redesign a module. Migrate to a new framework.

Each task teaches you something no book can: intuition for when to trust AI, when to push back, and when to take over.

Thank You

Writing a technical book in a field that moves this fast is an act of optimism — the optimism that principles outlast products and that good practices survive tool changes. I believe the approaches in this book will remain valuable long after the specific model names and version numbers are outdated.

Thank you for investing your time in becoming a better developer. The industry needs people who use AI thoughtfully, not just enthusiastically. People who care about code quality, user experience, and building things that work — and who see AI as a means to those ends, not an end in itself.

Now close this book and go build something.

— Ken Tannenbaum, 2026

Appendix A: AI Coding Tool Comparison

Side-by-Side Comparison of Major AI Coding Tools (2026)

This appendix compares the major AI coding tools available to developers as of early 2026. Capabilities change fast — check each tool's documentation for the latest features and pricing.

Quick Reference

For each tool: **type | base model | context | paid price**.

- **Claude Code** — CLI + IDE agent | Claude (Anthropic) | 1M tokens | $20–100/mo
- **GitHub Copilot** — IDE extension | GPT-5.4 / Claude (mixed) | Varies | $10–39/mo
- **Cursor** — AI-native IDE | Multiple (GPT-5.4, Claude, custom) | Varies by model | $20–40/mo
- **ChatGPT** — Chat + canvas | GPT-5.4 | 128K tokens | $20/mo
- **Gemini Code Assist** — IDE extension | Gemini 3.1 Pro/Flash | 1M tokens | $19/mo
- **Amazon Q** — IDE + CLI | Multiple (Claude, custom) | Varies | Free–$19/mo
- **Windsurf** — AI-native IDE | Multiple | Varies | $15–50/mo

Capability Comparison

- **Codebase awareness:** Full repo indexing — Claude Code, Cursor, Amazon Q, Windsurf. Repo-level — Copilot. Workspace-level — Gemini. Manual context — ChatGPT.
- **Agentic mode:** Native — Claude Code. Composer — Cursor. Cascade — Windsurf. Yes — Amazon Q. Workspace preview — Copilot. Limited — ChatGPT, Gemini.

- **Terminal integration:** Native — Claude Code, Amazon Q. Yes — Cursor, Windsurf. None — Copilot, ChatGPT, Gemini.
- **Multi-file editing:** Yes — Claude Code, Cursor, Amazon Q, Windsurf. Limited — Copilot, Gemini. None (copy/paste) — ChatGPT.
- **Free tier:** Yes — ChatGPT (mini), Gemini (limited), Amazon Q. Limited — all others.

Detailed Tool Profiles

Claude Code

What it is: A command-line tool and IDE extension from Anthropic that operates as a full development agent. You interact with it in your terminal or editor; it reads your codebase, writes code, runs commands, and creates commits.

Best for: - Developers who live in the terminal - Large-scale refactoring across many files - Complex tasks that require understanding the full codebase - Agentic workflows (plan, implement, test, iterate)

Strengths: - Massive context window (1M tokens) means it can hold your entire codebase in memory - True agentic behavior — it plans, executes, and iterates without hand-holding - Excellent at following project conventions when given a CLAUDE.md file - Strong at complex reasoning tasks (debugging, architecture decisions) - Native terminal integration means it can run your tests, check build output, and fix errors - Extended thinking mode for complex problems

Weaknesses: - Requires comfort with command-line workflows - Can be expensive at high usage (token-based pricing on some plans) - No visual UI for code diffs (works in terminal text) - Newer tool, smaller community than Copilot

Best plan for teams: Claude Max ($100/month) or Team plan for centralized billing and admin controls.

Setup: See Appendix C for configuration details.

GitHub Copilot

What it is: The most widely adopted AI coding assistant. Started as autocomplete, now includes chat, code review, and workspace features. Integrated into VS Code, JetBrains, Neovim, and GitHub.com.

Best for: - Inline code completion while typing - Developers who want AI assistance without changing their workflow - Teams already on GitHub Enterprise - Developers new to AI coding tools

Strengths: - Best-in-class autocomplete — predictions appear as you type, often reading your mind - Deep GitHub integration (PR summaries, code review, issue triage) - Massive ecosystem — most third-party tools and extensions support it - Available in nearly every editor - Business tier includes IP indemnification - Copilot Workspace for multi-file agentic tasks (evolving rapidly)

Weaknesses: - Chat quality can be inconsistent (model selection varies) - Autocomplete can be distracting for some developers - Less effective for large-scale cross-file changes compared to agentic tools - Context window is smaller than Claude Code's for whole-codebase understanding

Best plan for teams: Copilot Business ($19/month/user) for admin controls, audit logs, and IP indemnification. Enterprise ($39/month/user) adds fine-tuned models and security features.

Cursor

What it is: A fork of VS Code rebuilt around AI. Every feature is designed with AI assistance in mind. Includes inline editing, chat, and an agentic Composer mode for multi-file changes.

Best for: - Developers who want AI deeply integrated into their editor experience - Frontend and full-stack development - Rapid prototyping and iteration - Developers who prefer visual interfaces over terminal

Strengths: - Seamless integration — AI is part of the editor, not bolted on - Composer mode handles multi-file changes with a visual diff interface - Model flexibility — use Claude, GPT-5.4, or custom models - Codebase indexing for project-aware completions - Inline editing (Cmd+K) for quick targeted changes - Tab completion that understands multi-cursor editing

Weaknesses: - VS Code fork means you depend on Cursor's update cycle for VS Code features - Some VS Code extensions may not work perfectly - Premium models (Claude Opus 4.7, GPT-5.4) cost extra or have limited requests - Not available for JetBrains, Vim, or other editors

Best plan for teams: Business ($40/month/user) for team features and admin controls.

ChatGPT (with Code Interpreter / Canvas)

What it is: OpenAI's general-purpose AI assistant with code generation capabilities. Canvas mode provides a dedicated coding interface. Code Interpreter executes Python code.

Best for: - Exploratory coding and prototyping - Learning new languages or frameworks - One-off scripts and data analysis - Developers who prefer a conversational interface - Non-IDE coding tasks (architecture discussions, algorithm design)

Strengths: - Excellent general knowledge — great for "how do I do X in Y" questions - Canvas mode provides a focused coding environment - Code Interpreter executes code and shows results - Multimodal — paste screenshots of errors, diagrams, or UI mockups - Available on mobile for coding questions on the go

Weaknesses: - No direct IDE integration for inline assistance - No codebase awareness — you must paste code manually - Cannot run your tests or access your file system - Context management is manual (copy/paste relevant files) - Not designed for production development workflows

Best plan: Plus ($20/month) for GPT-5.4 access and higher rate limits.

Gemini Code Assist

What it is: Google's AI coding assistant, powered by Gemini models. Available as a VS Code and JetBrains extension, and integrated into Google Cloud Platform.

Best for: - Teams on Google Cloud Platform - Full-stack development with Google technologies (Firebase, Cloud Run, etc.) - Large file understanding (Gemini's 1M+ token context) - Multimodal tasks (understanding diagrams, screenshots, UI mockups)

Strengths: - Enormous context window — can process very large files and codebases - Strong Google Cloud integration (deploy, debug, monitor from the IDE) - Good multimodal support (paste images, diagrams) - Competitive free tier for individual developers - Code customization trained on your organization's code (Enterprise tier)

Weaknesses: - Smaller ecosystem of integrations compared to Copilot - Quality can be inconsistent for complex reasoning tasks - Agentic capabilities are less mature than Claude Code or Cursor - Less community content (tutorials, tips, extensions)

Best plan for teams: Enterprise (pricing varies) for code customization and Google Cloud integration.

Amazon Q Developer

What it is: AWS's AI coding assistant. Offers code suggestions, chat, and agentic capabilities, with deep AWS integration.

Best for: - Teams building on AWS - Infrastructure and DevOps work (CloudFormation, CDK, Terraform) - Java and Python development - Modernization and migration projects

Strengths: - Deep AWS integration — understands your AWS resources and can generate infrastructure code - Strong at Java and enterprise languages - Agent capabilities for code transformation (Java 8 to 17 migration, etc.) - Generous free tier - Security scanning built in - Can analyze and explain your AWS billing

Weaknesses: - AWS-centric — less useful if you are not on AWS - Fewer language/framework strengths outside Java/Python - Smaller community than Copilot or ChatGPT - IDE support is more limited than Copilot

Best plan for teams: Pro ($19/month/user) for higher limits and admin features.

Windsurf (by Codeium)

What it is: An AI-native IDE (VS Code fork) with Cascade — an agentic system that plans and executes multi-step development tasks.

Best for: - Developers who want a free or low-cost AI IDE - Agentic development workflows similar to Cursor - Teams evaluating alternatives to Cursor

Strengths: - Cascade agentic mode is capable and improving rapidly - Competitive pricing with a usable free tier - Codebase-aware completions and chat - Familiar VS Code interface - "Flows" feature for combining multiple AI actions

Weaknesses: - Newer product, still maturing - Smaller community and fewer learning resources - Some features lag behind Cursor's implementation - Enterprise features are still developing

Best plan for teams: Pro ($15–50/month) depending on usage needs.

Choosing the Right Tool

For Individual Developers

Start with **GitHub Copilot** for autocomplete and **Claude Code** for complex tasks. This combination covers the full spectrum, from quick inline suggestions to large-scale agentic work. Add ChatGPT for conversational exploration and learning.

For Teams

Standardize on **GitHub Copilot Business** as the baseline — everyone gets autocomplete, and it is the least disruptive to adopt. Add **Claude Code** or **Cursor** for developers doing complex, multi-file work. Use the team adoption approach from Chapter 23 to roll out.

For Enterprise

Choose based on your cloud provider relationship: - **AWS shop:** Amazon Q + Copilot - **Google Cloud shop:** Gemini Code Assist + Copilot - **Cloud-agnostic:** Copilot + Claude Code or Cursor

Decision Flowchart

```
Do you work primarily in a terminal?
├── Yes → Claude Code (primary) + Copilot (autocomplete)
└── No → Do you want AI built into your editor?
    ├── Yes → Cursor or Windsurf (primary) + Copilot (if needed)
    └── No → Copilot (primary) + ChatGPT (complex questions)

Are you on AWS/GCP?
├── AWS → Add Amazon Q for infrastructure work
└── GCP → Add Gemini Code Assist for cloud integration
```

Cost Comparison for a Team of 10

Configuration	Monthly Cost	What You Get
Copilot Business only	$190	Autocomplete + chat for everyone
Copilot + Claude Code Max (5 power users)	$690	Autocomplete for all + agentic for power users
Cursor Business for all	$400	AI-native IDE for everyone
Copilot + Cursor (5 power users)	$390	Autocomplete for all + AI IDE for power users
Full stack (Copilot all + Claude Code 5)	$690	Maximum coverage

All of these configurations pay for themselves in the first week of measurable productivity gains for a team of experienced developers. The question is not whether to invest but which combination fits your workflow.

A Note on Rapid Change

This comparison reflects the state of these tools as of early 2026. Every tool on this list ships major updates monthly. Features listed as weaknesses may be strengths by the time you read this. Check current documentation and community reviews before making purchasing decisions.

The principles from the rest of this book — how to use AI effectively, how to verify output, how to integrate AI into your workflow — apply regardless of which tool you choose.

Appendix B: Prompt Patterns for Developers

40+ Developer-Specific Prompt Patterns

These prompt patterns are templates you can adapt for your own projects. Each pattern includes the structure, an example, and notes on when to use it. They work across all major AI coding tools — Claude Code, Copilot Chat, Cursor, ChatGPT, and others.

The key principle: the more context and constraints you provide, the better the output. Vague prompts produce generic code. Specific prompts produce production-ready code.

Code Generation Patterns

1. The Constrained Generator

Structure: Describe what you want + list specific constraints and requirements.

Example:

```
Create a rate limiter middleware for Express that:
- Uses a sliding window algorithm (not fixed window)
- Stores state in Redis using ioredis
- Supports per-route and per-user limits
- Returns standard 429 responses with Retry-After header
- Includes TypeScript types for configuration
- Follows our middleware pattern in src/middleware/auth.ts
```

When to use: Any time you need generated code to meet specific technical requirements. The constraints prevent the AI from making its own architectural decisions.

2. The Pattern Follower

Structure: Point to an existing file and ask for code that follows the same pattern.

Example:

```
Look at src/routes/users.ts and create a new route file
src/routes/organizations.ts that follows the same patterns –
same validation approach, same error handling, same response
format. The organizations resource has: id, name, ownerId,
plan (free/pro/enterprise), createdAt, updatedAt.
```

When to use: When your project has established patterns and you want consistency. This is one of the most effective patterns for codebase-aware tools like Claude Code and Cursor.

3. The Incremental Builder

Structure: Break complex features into small, sequential steps.

Example:

```
I need to add WebSocket support to this Express app. Let's
do this step by step:

Step 1: Add the ws dependency and create a WebSocket server
that attaches to the existing HTTP server in src/server.ts.
Do only step 1 – I'll review before we continue.
```

When to use: For complex features where you want to review each piece before proceeding. Prevents the AI from building a castle on a flawed foundation.

4. The Signature-First Generator

Structure: Define the interface/types first, then ask for the implementation.

Example:

```
Implement this interface:

interface CacheService {
  get<T>(key: string): Promise<T | null>;
  set<T>(key: string, value: T, ttlSeconds?: number): Promise<void>;
  delete(key: string): Promise<void>;
```

```
  invalidatePattern(pattern: string): Promise<number>;
}

Use Redis as the backend. Handle connection errors gracefully —
cache failures should never crash the application. Log warnings
on cache misses and errors.
```

When to use: When you have a clear interface contract. The types constrain the AI's output to exactly what you need.

5. The Example-Driven Generator

Structure: Show input/output examples and ask for the implementation.

Example:

```
Write a function that parses our log format into structured objects.

Input:  "2026-03-15T10:23:45Z [ERROR] auth-service req=abc123 | Failed to valida
te token: expired"
Output: { timestamp: "2026-03-15T10:23:45Z", level: "ERROR", service: "auth-serv
ice", requestId: "abc123", message: "Failed to validate token: expired" }

Input:  "2026-03-15T10:23:46Z [INFO] api-gateway req=def456 | Request completed
in 234ms"
Output: { timestamp: "2026-03-15T10:23:46Z", level: "INFO", service: "api-gatewa
y", requestId: "def456", message: "Request completed in 234ms" }

Handle malformed lines by returning null. Include TypeScript types.
```

When to use: When the specification is clearest as examples rather than prose. Particularly effective for parsers, formatters, and data transformation functions.

Testing Patterns

6. The Coverage Gap Filler

Structure: Point to code and ask for tests targeting untested paths.

Example:

```
Look at src/services/payment.ts and its existing tests in
src/services/payment.test.ts. Write tests for the code paths
that are NOT currently covered. Focus on error handling paths,
edge cases, and boundary conditions.
```

7. The Edge Case Hunter

Structure: Ask explicitly for edge case tests.

Example:

```
Write tests for the validateEmail function in src/utils/validation.ts.
Include edge cases:
- Empty string, null, undefined
- Strings with unicode characters
- Very long email addresses (>254 chars)
- Multiple @ symbols
- Valid but unusual formats (plus addressing, subdomains)
- Strings that look almost like emails but are not
```

8. The Test-First Specifier

Structure: Write tests before implementation, using tests as specification.

Example:

```
Write Jest tests for a ShoppingCart class that does NOT exist yet.
The tests should specify this behavior:
- Adding items with quantity
- Removing items
- Updating quantities (including to zero, which removes the item)
- Calculating subtotal, tax (8.25%), and total
- Applying percentage and fixed-amount discount codes
- Maximum one discount code at a time (applying a new one replaces the old one)
- Serialization to/from JSON

Write the tests only. I will implement the class to make them pass.
```

9. The Integration Test Scaffolder

Structure: Describe the system interactions and ask for integration tests.

Example:

```
Write integration tests for the POST /api/orders endpoint.
The test should:
- Set up a test database with seed data (use our test helpers in tests/helpers/)
- Create an authenticated user session
- Test the full flow: create order → verify database state → verify
  event was published to the order.created queue
- Clean up after each test
- Use our existing test patterns in tests/integration/
```

10. The Property-Based Test Writer

Structure: Describe invariants and let the AI generate property tests.

Example:

```
Write fast-check property-based tests for the money arithmetic
functions in src/utils/money.ts. Test these properties:
- Addition is commutative: add(a, b) === add(b, a)
- Addition is associative: add(add(a, b), c) === add(a, add(b, c))
- Multiplication distributes: multiply(add(a, b), n) === add(multiply(a, n), mul
tiply(b, n))
- Rounding never loses more than 1 cent
- Currency conversion roundtrip is within 1% of original
```

Debugging Patterns

11. The Error Explainer

Structure: Paste the error and relevant code, ask for diagnosis.

Example:

```
This error occurs intermittently in production (about 5% of requests):

Error: ECONNRESET
    at TLSWrap.onStreamRead (node:internal/stream_base_commons:217:20)
    at TLSWrap.callbackTrampoline (node:internal/async_hooks:128:17)

Here is the code that makes the outbound request: [paste code]
Here is our connection pooling config: [paste config]

What are the likely causes and how do I fix each one?
```

12. The Rubber Duck Debugger

Structure: Explain the bug in detail, including what you have already tried.

Example:

```
I have a bug where user sessions expire after exactly 15 minutes
even though our session TTL is set to 24 hours. Here is what I know:
- Redis TTL on session keys shows 86400 (correct)
- The Express session middleware config sets maxAge to 86400000ms
- It only happens for users going through our load balancer
- Direct connections to a single server work fine

I have already checked: cookie domain settings, Redis cluster sync,
and session middleware order. What else could cause this?
```

13. The Log Analyzer

Structure: Paste logs and ask for pattern identification.

Example:

```
These are the last 50 lines of our application log during the
outage. Identify the sequence of events that led to the failure,
the root cause, and what we should fix:

[paste logs]
```

14. The Performance Profiler

Structure: Share performance data and ask for optimization recommendations.

Example:

```
This database query takes 3.2 seconds on our production dataset
(2M rows in orders, 500K in users). Here is the query, the EXPLAIN
ANALYZE output, and the current indexes:

[paste query]
[paste explain output]
[paste index definitions]

What specific changes will bring this under 100ms?
```

Refactoring Patterns

15. The Extract and Restructure

Structure: Identify what to extract and where it should go.

Example:

```
The function processOrder in src/services/orders.ts is 200 lines
long and does too many things. Extract it into separate functions:
- validateOrder (input validation)
- calculatePricing (pricing logic)
- processPayment (payment gateway interaction)
- createOrderRecord (database insertion)
- sendConfirmation (notification)

Keep them in the same file for now. Each function should be
independently testable. The main processOrder function should
orchestrate them.
```

16. The Pattern Migrator

Structure: Show the old pattern and the desired new pattern.

Example:

```
We are migrating from callback-style error handling to Result types.

Old pattern (current):
function getUser(id: string, callback: (err: Error | null, user?: User) => void)

New pattern (desired):
function getUser(id: string): Promise<Result<User, AppError>>

Convert all functions in src/services/user-service.ts from the
old pattern to the new pattern. Use the Result type defined in
src/types/result.ts.
```

17. The Dependency Updater

Structure: Specify the migration and ask for all necessary changes.

Example:

```
Migrate this file from Express 4 to Express 5. Here are the
breaking changes I know about:
- req.query returns URLSearchParams
- Path route matching is stricter
- Removed app.del() (use app.delete())
- Promise-based middleware error handling

Update the code and flag anything I might have missed.
```

18. The Dead Code Finder

Structure: Ask AI to identify unused code.

Example:

```
Analyze src/utils/ and identify functions that are not imported or
used anywhere in the codebase. For each function you identify,
confirm it is truly unused by searching for all references.
List them so I can decide what to remove.
```

Documentation Patterns

19. The API Documenter

Structure: Point to endpoints and request comprehensive API docs.

Example:

```
Generate OpenAPI 3.1 documentation for all endpoints in
src/routes/. Include:
- Request/response schemas (derived from our TypeScript types)
- Example requests and responses
- Error response formats
- Authentication requirements
- Rate limit information
```

20. The README Generator

Structure: Describe the project and what the README should cover.

Example:

```
Write a README for this project based on the codebase. Include:
- What the project does (one paragraph)
- Prerequisites and setup instructions
- How to run locally
- How to run tests
- Project structure overview
- Environment variables (from .env.example)
- Contributing guidelines (brief)
```

21. The Architecture Decision Record

Structure: Describe the decision context and ask for a structured ADR.

Example:

```
Write an ADR for our decision to switch from REST to GraphQL for
the mobile API. Context: our mobile app makes 8-12 REST calls per
screen, latency on cellular is killing UX, and we need field-level
selection. We considered: BFF pattern, GraphQL, and tRPC. We chose
GraphQL. Use the standard ADR format (Title, Status, Context,
Decision, Consequences).
```

22. The Inline Commenter

Structure: Ask for comments that explain "why," not "what."

Example:

```
Add inline comments to this function. Do not comment obvious things
(what the code does is visible). Comment WHY — business reasons,
non-obvious constraints, workarounds, and anything a new developer
would find confusing. If something does not need a comment, do not
add one.
```

Architecture Patterns

23. The Trade-Off Analyzer

Structure: Present options and ask for structured analysis.

Example:

```
We need to add full-text search to our application. Compare these options:
1. PostgreSQL full-text search (pg_trgm + tsvector)
2. Elasticsearch
3. Typesense
4. Meilisearch

Our context: 2M documents, 50 searches/second peak, team has
strong PostgreSQL skills but no Elasticsearch experience, we run
on AWS, budget is a concern.

For each option: estimated cost, operational complexity, search
quality, and your recommendation with reasoning.
```

24. The System Design Reviewer

Structure: Present your design and ask for critique.

Example:

```
Here is our planned architecture for the notification service:
[paste diagram or description]

Review this design for:
- Single points of failure
- Scaling bottlenecks
- Data consistency issues
- Operational complexity
- Missing error handling paths

Be specific — tell me what will break and under what conditions.
```

25. The Schema Designer

Structure: Describe the domain and ask for a schema.

Example:

```
Design a PostgreSQL schema for a multi-tenant SaaS project
management tool. Entities: organizations, users, projects,
tasks, comments, attachments. Requirements:
- Row-level security for tenant isolation
- Soft deletes on all entities
- Audit trail (who changed what, when)
- Support for custom fields on tasks
- Efficient queries for "my tasks across all projects"
```

DevOps Patterns

26. The CI Pipeline Builder

Structure: Describe your stack and deployment requirements.

Example:

```
Write a GitHub Actions workflow for a Node.js monorepo using pnpm
workspaces. It should:
- Run on push to main and on PRs
- Install dependencies with caching
- Run lint, type check, and tests in parallel for changed packages only
- Build only affected packages
- Deploy to staging on main push, production on release tags
- Notify Slack on failure
```

27. The Dockerfile Optimizer

Structure: Share your Dockerfile and ask for optimization.

Example:

```
Optimize this Dockerfile for production. Current image is 1.2GB
and takes 8 minutes to build. Goals:
- Minimize image size (target: under 200MB)
- Maximize layer caching for fast rebuilds
- Multi-stage build
- Non-root user
- Health check
- No dev dependencies in final image

[paste Dockerfile]
```

28. The Infrastructure Describer

Structure: Describe desired infrastructure in plain English.

Example:

```
Write Terraform for an AWS setup:
- ECS Fargate cluster running our API (2 tasks min, 10 max)
- ALB with HTTPS termination (cert already in ACM)
```

```
- RDS PostgreSQL 16 (db.r6g.large, multi-AZ)
- ElastiCache Redis cluster (cache.r6g.large, 2 nodes)
- VPC with public/private subnets across 3 AZs
- Security groups: ALB public, ECS and RDS private only

Use modules and variables. Include outputs for the ALB DNS
name and RDS endpoint.
```

Security Patterns

29. The Vulnerability Scanner

Structure: Ask for a security review of specific code.

Example:

```
Review this authentication flow for security vulnerabilities:
[paste code]

Check for: injection attacks, timing attacks, session fixation,
CSRF, insecure token storage, missing rate limiting, information
leakage in error messages, and anything else you see.
```

30. The Secure Implementation

Structure: Specify security requirements upfront.

Example:

```
Implement a password reset flow. Security requirements:
- Token: cryptographically random, 256-bit, single-use
- Expiration: 1 hour
- Rate limit: 3 requests per email per hour
- Do not reveal whether the email exists in our system
- Log all reset attempts for audit
- Invalidate all existing tokens when a new one is generated
- Use constant-time comparison for token validation
```

Database Patterns

31. The Query Optimizer

Structure: Share the slow query and ask for optimization.

Example:

```
This query runs in our reporting dashboard and takes 12 seconds:
[paste query]
[paste EXPLAIN ANALYZE output]
[paste relevant table schemas and indexes]

Optimize it. Consider: query rewriting, index changes,
materialized views, or query decomposition. Show me the
improved version with expected performance.
```

32. The Migration Writer

Structure: Describe the schema change with safety requirements.

Example:

```
Write a database migration to add a 'status' column to the
orders table. Requirements:
- Default value: 'pending'
- Must be backwards compatible (old code still works)
- Zero-downtime deployment (no table locks on a 50M row table)
- Include both up and down migrations
- Add an index for filtering by status
- Use our migration format (see migrations/001_example.sql)
```

Frontend Patterns

33. The Component Generator

Structure: Describe the component with specific UI/UX requirements.

Example:

```
Create a React data table component with:
- Server-side pagination, sorting, and filtering
```

```
- Column definitions passed as props (type-safe with generics)
- Row selection (single and multi)
- Loading and empty states
- Responsive (stacks on mobile)
- Uses our design system components from @/components/ui
- Follows the patterns in src/components/UserTable.tsx
```

34. The State Machine Designer

Structure: Describe the states and transitions.

Example:

```
Create an XState machine for our checkout flow:
States: cart → shipping → payment → review → processing → confirmed/failed
Transitions:
- cart → shipping: when cart is non-empty
- shipping → payment: when address is validated
- payment → review: when payment method is saved
- review → processing: on submit
- processing → confirmed: on payment success
- processing → failed: on payment failure
- failed → payment: on retry
- Any state → cart: on cancel

Include context for: cart items, shipping address, payment
method, order ID, and error message.
```

35. The Accessibility Auditor

Structure: Ask for an accessibility review of a component.

Example:

```
Review this React component for accessibility issues:
[paste component]

Check for: keyboard navigation, screen reader support, ARIA
labels, color contrast, focus management, and WCAG 2.1 AA
compliance. Provide the fixed version.
```

Code Review Patterns

36. The Pre-Review Self-Check

Structure: Ask AI to review your code before you submit the PR.

Example:

```
I am about to submit this as a PR. Review it as if you were
a senior developer on my team. Be critical. Check for:
- Bugs and logic errors
- Missing error handling
- Performance issues
- Security concerns
- Naming and readability
- Missing tests
- Anything that will cause problems in production

[paste diff or point to files]
```

37. The Convention Enforcer

Structure: Check code against your team's standards.

Example:

```
Check this code against our team conventions:
- Functions under 30 lines
- No nested callbacks deeper than 2 levels
- All public functions have JSDoc comments
- Error types extend AppError
- Database queries use parameterized queries only
- No magic numbers (use named constants)

List every violation with line numbers.
```

Learning and Exploration Patterns

38. The Concept Explainer

Structure: Ask for an explanation at your level of expertise.

Example:

```
Explain Raft consensus algorithm to me. I understand distributed
systems basics (CAP theorem, eventual consistency) but have never
implemented a consensus protocol. Use a concrete example with
3 nodes. Show me the leader election and log replication flow
step by step.
```

39. The Codebase Explorer

Structure: Ask the AI to explain unfamiliar code.

Example:

```
I just joined this project and need to understand the authentication
system. Walk me through:
1. How a login request flows from the API endpoint to the database
2. How tokens are generated and validated
3. How middleware protects authenticated routes
4. Where the session/token configuration lives

Point me to the specific files and functions for each step.
```

40. The Technology Evaluator

Structure: Ask for an honest assessment of a technology choice.

Example:

```
We are considering adopting tRPC to replace our REST API layer.
Our team: 6 developers, strong TypeScript skills, monorepo with
shared types. Current stack: Express + OpenAPI + code-generated
clients.

Give me the honest case for and against. Include: migration cost,
learning curve, long-term maintainability, and what we would gain
and lose. Do not be a cheerleader – I want the real trade-offs.
```

Meta Patterns

41. The Context Setter

Structure: Provide project context before any request.

Example:

```
Context for this session: I am working on a B2B SaaS invoicing
application. Stack: Next.js 15, Prisma, PostgreSQL, Stripe for
payments. Monorepo with pnpm workspaces. We follow domain-driven
design with services in src/domains/. Tests use Vitest.

Keep this context in mind for all requests in this conversation.
```

42. The Output Format Controller

Structure: Specify exactly how you want the response structured.

Example:

```
[your request]

Format your response as:
1. Summary (2-3 sentences)
2. Code changes (with file paths, full functions, not snippets)
3. Test changes (same format)
4. Migration steps (if any database changes)
5. Things I should manually verify
```

43. The Iterative Refiner

Structure: Start broad, then narrow.

Example:

```
Round 1: "Give me three approaches for implementing real-time
notifications in our stack."

Round 2: "Expand on option 2 (WebSocket with Redis pub/sub).
Show me the architecture."

Round 3: "Implement the WebSocket server module. Follow our
patterns in src/services/."

Round 4: "Now write the client-side hook that connects to this
WebSocket server."
```

44. The Constraint Tightener

Structure: Add constraints when the first output is too generic.

Example:

```
First attempt: "Write a caching layer for our API."
→ Output is too generic.

Refined: "Write a caching layer for our Express API that:
- Caches GET responses in Redis
- Uses URL + auth token hash as cache key
- TTL varies by route (configurable per-route)
- Invalidates on POST/PUT/DELETE to the same resource
- Bypasses cache when Cache-Control: no-cache is sent
- Returns X-Cache: HIT/MISS header
- Uses our existing Redis connection from src/lib/redis.ts"
```

Tips for Better Prompts

1. **Be specific about the "how," not just the "what."** "Write a login form" produces generic code. "Write a login form using React Hook Form with Zod validation" produces usable code.

2. **Reference existing code.** "Follow the pattern in src/routes/users.ts" is worth a thousand words of description.

3. **Specify what you do not want.** "Do not use any/unknown types" or "Do not use raw SQL" prevents common AI defaults that violate your standards.

4. **Include error handling requirements.** AI often generates happy-path code unless you explicitly ask for error handling.

5. **Ask for tests alongside code.** "Write the function and its unit tests" produces better code than asking for each separately, because the AI considers testability during generation.

6. **State your experience level.** "Explain this to a senior developer who is new to Rust" gets a different (better) response than an unqualified "explain this."

7. **Iterate, do not restart.** If the output is 80% right, refine it. "Good, but change X and Y" is more effective than starting over with a new prompt.

8. **Provide examples of what good looks like.** Show a well-written function from your codebase and say “write the new function in this style.”

Appendix C: Configuration Guides

Setup Guides for Each Major AI Coding Tool

This appendix walks you through installing and configuring each major AI coding tool covered in this book. For each tool, you get installation steps, essential configuration, recommended settings, and project-level setup.

Claude Code

Installation

Claude Code runs as a CLI tool and integrates with your terminal and editors.

```
## Install via npm (requires Node.js 18+)
npm install -g @anthropic-ai/claude-code

## Verify installation
claude --version

## First run — authenticates with your Anthropic account
claude
```

Authentication

Claude Code authenticates through your Anthropic account. On first run, it opens a browser for OAuth, and your session persists after authentication.

For team deployments, set the API key directly:

```
## Environment variable (for CI/CD or scripted usage)
export ANTHROPIC_API_KEY=sk-ant-...
```

Project Configuration: CLAUDE.md

Create a `CLAUDE.md` file in your project root. Claude Code reads it automatically and uses it as persistent context for every interaction.

```
## CLAUDE.md

### Project Overview
This is a Node.js/TypeScript e-commerce API using Express,
Prisma, and PostgreSQL.

### Tech Stack
- Runtime: Node.js 22
- Language: TypeScript 5.7 (strict mode)
- Framework: Express 5
- ORM: Prisma 6
- Database: PostgreSQL 16
- Testing: Vitest
- Package manager: pnpm

### Conventions
- All API responses use the ApiResponse<T> wrapper (src/types/api.ts)
- Route handlers delegate to service functions (never query DB in routes)
- Tests live next to source files: foo.ts → foo.test.ts
- Use named exports, not default exports
- Error classes extend AppError (src/errors/base.ts)

### Commands
- Build: pnpm build
- Test: pnpm test
- Lint: pnpm lint
- Dev server: pnpm dev

### Do Not
- Use any/unknown without explicit justification
- Write raw SQL (use Prisma)
- Import internal packages with relative paths crossing package boundaries
- Skip input validation on API endpoints
```

Recommended Settings

Configure Claude Code behavior in `~/.claude/settings.json`:

```
{
  "permissions": {
    "allow": [
      "Read",
      "Glob",
      "Grep",
      "Bash(npm test*)",
```

```
      "Bash(pnpm *)",
      "Bash(git status)",
      "Bash(git diff*)",
      "Bash(git log*)"
    ],
    "deny": [
      "Bash(rm -rf *)",
      "Bash(git push --force*)",
      "Bash(git reset --hard*)"
    ]
  }
}
```

Usage Tips

- Use `/init` to generate a starter CLAUDE.md based on your project
- Press Escape to interrupt long-running operations
- Use `claude --resume` to continue a previous session
- In VS Code, use the Claude Code extension for integrated panel access
- For large tasks, use extended thinking: Claude shows its reasoning process

---

GitHub Copilot

Installation

VS Code

1. Open VS Code
2. Go to Extensions (Ctrl+Shift+X / Cmd+Shift+X)
3. Search "GitHub Copilot"
4. Install both "GitHub Copilot" and "GitHub Copilot Chat"
5. Sign in with your GitHub account when prompted

JetBrains (IntelliJ, WebStorm, PyCharm, etc.)

1. Open Settings → Plugins
2. Search "GitHub Copilot"
3. Install and restart the IDE
4. Sign in via Settings → GitHub Copilot

NEOVIM

```lua
-- Using lazy.nvim
{
  "zbirenbaum/copilot.lua",
  cmd = "Copilot",
  event = "InsertEnter",
  config = function()
    require("copilot").setup({
      suggestion = {
        enabled = true,
        auto_trigger = true,
        keymap = {
          accept = "<Tab>",
          next = "<M-]>",
          prev = "<M-[>",
          dismiss = "<C-]>",
        },
      },
    })
  end,
}
```

Configuration (VS Code)

Add to your `settings.json`:

```json
{
  "github.copilot.enable": {
    "*": true,
    "plaintext": false,
    "markdown": true,
    "yaml": true
  },
  "github.copilot.advanced": {
    "length": 500,
    "inlineSuggestCount": 3
  }
}
```

Project Configuration

Create `.github/copilot-instructions.md` in your repo:

```
## Copilot Instructions

When generating code for this project:
- Use TypeScript strict mode
- Follow the patterns in src/routes/ for API endpoints
- Use Prisma for all database queries
- Wrap responses in ApiResponse<T>
- Include error handling for all async operations
- Write Vitest tests for all new functions
```

Customizing Autocomplete Behavior

Copilot's autocomplete is configurable:

- **Tab to accept:** The default. Press Tab to accept the suggestion.
- **Accept word by word:** Ctrl+Right (Cmd+Right on Mac) accepts one word at a time.
- **Cycle suggestions:** Alt+] and Alt+[(Option+] and Option+[) cycle through alternatives.
- **Dismiss:** Press Escape to dismiss the current suggestion.

Copilot Chat Slash Commands

Use these in the Copilot Chat panel:

Command	What it does
/explain	Explains selected code
/fix	Suggests a fix for problems in selected code
/tests	Generates tests for selected code
/doc	Generates documentation for selected code
/optimize	Suggests performance improvements
/new	Scaffolds a new project or file

Cursor

Installation

1. Download from cursor.com

2. Install (available for macOS, Windows, Linux)
3. Import your VS Code settings when prompted (optional but recommended)
4. Sign in and choose your plan

Key Configuration

Open Settings (Ctrl+, / Cmd+,) and search for Cursor-specific settings:

```
{
  "cursor.cpp.enablePartialAccepts": true,
  "cursor.general.enableShadowWorkspace": true,
  "cursor.chat.defaultModel": "claude-sonnet-4-5",
  "cursor.tab.useRecentFiles": true
}
```

Project Configuration: .cursorrules

Create a `.cursorrules` file in your project root:

```
You are an expert TypeScript developer working on a Node.js
e-commerce API.

Tech stack: Express 5, Prisma 6, PostgreSQL, Vitest.

Rules:
- Always use strict TypeScript (no any/unknown without justification)
- API responses use ApiResponse<T> from src/types/api.ts
- Database access only through Prisma (no raw SQL)
- Every new function needs a corresponding test
- Error handling uses AppError hierarchy from src/errors/
- Follow existing patterns in the codebase

When editing files, preserve the existing code style (indentation,
quote style, semicolon usage).
```

Essential Keyboard Shortcuts

Action	macOS	Windows/Linux
Inline edit	Cmd+K	Ctrl+K
Open chat	Cmd+L	Ctrl+L
Open Composer	Cmd+I	Ctrl+I

Action	macOS	Windows/Linux
Accept suggestion	Tab	Tab
Toggle AI panel	Cmd+Shift+A	Ctrl+Shift+A
Reference file in chat	@file	@file
Reference codebase	@codebase	@codebase

Composer (Agentic Mode)

Cursor's Composer mode handles multi-file changes:

1. Open Composer with Cmd+I (Ctrl+I)
2. Describe the change you want across your project
3. Composer proposes changes to multiple files
4. Review the diff for each file
5. Accept or reject individual file changes

Best practices for Composer: - Be specific about which files to modify - Use @file references to point to relevant code - Review every changed file — never bulk-accept without reading - For large changes, break them into smaller Composer requests

---

ChatGPT

Setup

1. Go to chat.openai.com or install the desktop app
2. Subscribe to Plus ($20/month) for GPT-5.4 access
3. Enable "Code Interpreter" in your chat settings for executable Python

Configuration for Coding

Set up a custom instruction (Settings → Personalization → Custom Instructions):

What would you like ChatGPT to know about you?

```
I am a senior full-stack developer working primarily with
TypeScript, Node.js, React, and PostgreSQL. I use pnpm,
```

```
Vitest for testing, and Prisma as my ORM. I prefer concise,
production-quality code over explanatory examples.
```

How would you like ChatGPT to respond?

```
- Provide complete, runnable code (not snippets with "..." placeholders)
- Use TypeScript strict mode
- Include error handling
- Show file paths for where code should go
- Skip obvious explanations – focus on the non-obvious parts
- When I ask for a review, be critical – do not just praise the code
```

Canvas Mode

For focused coding work, use Canvas mode: 1. Start a new conversation 2. Ask for a coding task 3. Click "Open in Canvas" when the code appears 4. Edit directly in the canvas with AI assistance 5. Use the canvas toolbar to explain, fix bugs, add comments, or port to another language

Code Interpreter Tips

Code Interpreter runs Python in a sandboxed environment. Use it for:

- Data analysis and visualization
- Testing algorithms before implementing in your stack
- File format conversions
- Quick prototyping with immediate execution
- Debugging complex logic step by step

```
Upload a CSV and say: "Analyze this data. Show me the
distribution of response times, identify outliers, and
suggest what might be causing the slowest 5% of requests."
```

Gemini Code Assist

Installation

VS Code

1. Install the "Gemini Code Assist" extension from the marketplace
2. Sign in with your Google account

3. Select your Google Cloud project (if using Enterprise features)

JetBrains

1. Install the "Gemini Code Assist" plugin
2. Sign in with your Google account
3. Configure your Cloud project in Settings → Gemini

Configuration

VS Code settings:

```
{
  "gemini.codeAssist.enableInlineCompletion": true,
  "gemini.codeAssist.enableChat": true,
  "gemini.codeAssist.model": "gemini-pro"
}
```

Google Cloud Integration

On Google Cloud, Gemini integrates with your infrastructure:

```
## Install the gcloud CLI Gemini component
gcloud components install gemini

## Use Gemini for Cloud-specific tasks
gcloud gemini code-assist generate \
  --prompt "Create a Cloud Run service config for a Node.js API"
```

Enterprise Code Customization

On the Enterprise tier, Gemini can be trained on your organization's code:

1. Go to Google Cloud Console → Gemini Code Assist
2. Configure code customization
3. Point to your repositories
4. Gemini indexes your code and uses it for context-aware suggestions

---

Amazon Q Developer

Installation

VS Code

1. Install "Amazon Q" extension from the marketplace
2. Sign in with AWS Builder ID (free) or AWS IAM Identity Center (Pro)

JetBrains

1. Install "Amazon Q" plugin
2. Sign in with your AWS credentials

CLI

```
## The Q CLI is included in AWS CLI v2.15+
aws q developer chat --prompt "How do I set up a VPC with private subnets?"
```

Configuration

VS Code settings:

```
{
  "amazonQ.shareContentWithAWS": false,
  "amazonQ.telemetry": false
}
```

AWS-Specific Features

Amazon Q shines for AWS infrastructure work:

```
"Generate a CDK stack for a Fargate service behind an ALB,
with an RDS PostgreSQL database in a private subnet, and
a Redis ElastiCache cluster. Include security groups that
follow least-privilege principles."
```

Q can also: - Explain your AWS bill and suggest cost savings - Analyze CloudWatch logs for errors - Generate IAM policies from plain English descriptions - Transform Java 8 code to Java 17 (automated migration)

Windsurf

Installation

1. Download from windsurf.com (formerly Codeium)
2. Install (available for macOS, Windows, Linux)
3. Import VS Code settings when prompted
4. Sign in and choose your plan

Configuration

Windsurf uses a `.windsurfrules` file for project-level configuration:

```
This project is a TypeScript/Express API with Prisma ORM.

When generating code:
- Use strict TypeScript
- Follow patterns in existing route handlers
- Include error handling and input validation
- Write tests using Vitest
- Use named exports
```

Cascade (Agentic Mode)

Windsurf's Cascade is its agentic feature:

1. Open the AI panel
2. Describe your task
3. Cascade reads relevant files, plans changes, and proposes edits
4. Review and accept changes file by file

Key Differences from Cursor

- Cascade is more autonomous — it reads more files proactively
- "Flows" feature chains multiple AI actions together
- Free tier is more generous for basic features
- Smaller extension ecosystem

General Tips for All Tools

Environment Variables

Never put API keys in your code. Use environment files instead:

```
## .env (never committed - add to .gitignore)
ANTHROPIC_API_KEY=sk-ant-...
OPENAI_API_KEY=sk-...
GITHUB_TOKEN=ghp_...
```

Keeping Tools Updated

AI tools update frequently. Set a monthly reminder to:

```
## Claude Code
npm update -g @anthropic-ai/claude-code

## Check for IDE extension updates
## VS Code: Extensions → check for updates
## JetBrains: Settings → Plugins → check for updates
```

Using Multiple Tools Together

The most effective setup combines different tools for different purposes:

Purpose	Tool
Inline autocomplete	GitHub Copilot
Complex multi-file tasks	Claude Code or Cursor Composer
Quick questions	Copilot Chat or Cursor Chat
Architecture exploration	ChatGPT or Claude
AWS infrastructure	Amazon Q
Google Cloud work	Gemini Code Assist

You do not have to pick one. Use the right tool for the task.

Appendix D: Glossary

Developer-Focused AI Terms

This glossary covers AI and machine learning terms as they relate to software development. Definitions are practical, not academic — focused on what the term means for your daily work.

Agent / Agentic AI: An AI system that can take actions autonomously — reading files, running commands, making API calls, and iterating on its own output. Claude Code's ability to read your codebase, write code, run tests, and fix failures without step-by-step instructions is an example of agentic behavior. Contrast with a simple chat interface where you copy and paste code back and forth.

Attention Mechanism: The core innovation behind transformer models. Attention allows the model to focus on the most relevant parts of the input when generating each token. When you paste a 500-line file and ask about line 347, attention is what lets the model focus on that line and its surrounding context.

Autocomplete (AI-Powered): Real-time code suggestions that appear as you type, generated by an AI model that predicts what you are likely to write next. GitHub Copilot's inline suggestions are the canonical example. Not the same as traditional IDE autocomplete, which only suggests method names and variables — AI autocomplete generates entire blocks of logic.

Benchmark: A standardized test for measuring model performance. HumanEval (code generation), SWE-bench (real GitHub issues), and MMLU (general knowledge) are common benchmarks. Useful for comparing models but do not always reflect real-world coding performance.

Chain of Thought (CoT): A prompting technique where you ask the model to show its reasoning step by step before giving a final answer. Improves accuracy on complex tasks. Example: "Think through the possible race conditions in this code step by step before suggesting a fix."

Chunking: Breaking a large document or codebase into smaller pieces for processing. Relevant when working with embeddings or RAG systems — you chunk your documentation into sections so each can be independently embedded and retrieved.

Code Completion: See Autocomplete.

Completion (API): A single AI-generated response to a prompt. When you call the Anthropic or OpenAI API, you send a prompt and receive a completion. The term comes from the original concept of "completing" a text sequence.

Context Window: The maximum amount of text (measured in tokens) that a model can process in a single interaction. A 200K token context window can hold roughly 150,000 words or an entire medium-sized codebase. Larger context windows allow AI tools to understand more of your project at once.

Embedding: A numerical representation of text as a vector (array of numbers). Texts with similar meanings produce similar vectors. Used for semantic search, recommendation systems, and clustering. When you search your codebase by meaning rather than keywords, embeddings are the underlying technology.

Extended Thinking: A mode where the AI model spends additional computation reasoning about a problem before responding. Produces better results for complex tasks like debugging, architecture design, and multi-step reasoning. Available in Claude Code and other tools.

Few-Shot Prompting: Providing the model with a few examples of the desired input/output before asking your question. "Here are three examples of how we format error messages: examples. Now format this new error message the same way." Contrast with zero-shot (no examples) and one-shot (one example).

Fine-Tuning: Training an existing model on your specific data to specialize it for your use case. If you fine-tune a model on your company's codebase, it learns your patterns and conventions. Expensive and complex — most developers use prompt engineering and RAG instead.

Foundation Model: A large AI model trained on broad data that serves as the base for various applications. GPT-5.4, Claude, and Gemini are foundation models. They are general-purpose — specific applications (like coding assistants) are built on top of them.

Function Calling / Tool Use: The ability of an AI model to invoke external functions or tools as part of generating a response. When Claude Code runs your test suite to check if generated code works, it is using tool use. The model decides which tool to call, with what arguments, and incorporates the result into its response.

Grounding: Connecting AI responses to verified, factual sources. RAG is a grounding technique — instead of relying on the model's training data, you ground responses in your actual documentation or codebase.

Guardrails: Constraints placed on AI behavior to prevent unwanted outputs. System prompts that say "never reveal internal system details" or content filters that block harmful content are guardrails. Essential when exposing AI to end users.

Hallucination: When an AI model generates information that sounds plausible but is factually incorrect. In coding, this often manifests as references to APIs that do not exist, function signatures that are wrong, or libraries that were never published. The model is not "lying" — it is pattern-matching and the pattern led somewhere wrong.

Inference: The process of running a trained model to generate output. When you send a prompt to Claude and receive a response, that is inference. Inference costs (compute time, API fees) are the main ongoing expense of using AI tools.

In-Context Learning: A model's ability to learn from examples provided in the prompt, without any training or fine-tuning. When you show an AI three examples of your code style and it follows the pattern for new code, that is in-context learning.

Large Language Model (LLM): An AI model trained on vast amounts of text data that can generate, analyze, and transform text. GPT 5.4, Claude, Gemini, and Llama are LLMs. Despite the name, modern LLMs handle code, structured data, and multimodal inputs (images, audio) in addition to natural language.

Latency: The time between sending a request to an AI model and receiving the first response token. For interactive coding tools, latency matters — a 5-second delay breaks your flow. Smaller models (Claude Haiku, GPT-5.4 mini) have lower latency than larger models.

Model (AI): The trained neural network that generates responses. When people say "GPT-5.4" or "Claude Sonnet," they are referring to specific models. Different models have different capabilities, costs, and speed characteristics.

Multimodal: A model that can process multiple types of input — text, images, audio, video. Relevant for development when you want to paste a screenshot of a UI bug, a diagram of your architecture, or an image of a whiteboard sketch and have the AI understand it.

Parameter: In the context of model size, a parameter is a single learnable value in the neural network. "A 70B parameter model" has 70 billion parameters. More parameters generally means more capable but slower and more expensive. Not to be confused with function parameters in code.

Prompt: The input you provide to an AI model. In coding contexts, this includes your question, any code you paste, system instructions, and conversation history. Prompt quality is the single biggest determinant of output quality.

Prompt Engineering: The practice of crafting prompts to get better results from AI models. In development, this means writing specific, constrained requests rather than vague ones. See Appendix B for patterns.

RAG (Retrieval-Augmented Generation): A technique that combines search with AI generation. Instead of asking the model to answer from its training data, you first search your own data for relevant information, then include that information in the prompt. Used for building AI features that answer questions about your documentation, products, or knowledge base.

Rate Limiting (AI APIs): Usage limits imposed by AI providers — requests per minute, tokens per minute, or tokens per day. Exceeding rate limits returns 429 errors. Production applications must handle rate limiting with exponential backoff and queuing.

Reasoning Model: A model specifically optimized for multi-step logical reasoning, often using extended thinking or chain-of-thought internally. These models trade speed for accuracy on complex problems. Useful for debugging, architecture design, and code review.

Retrieval: The process of finding relevant information from a data store. In RAG systems, retrieval finds the documents most relevant to a user's query.

Vector similarity search using embeddings is the most common retrieval method for AI applications.

Semantic Search: Search based on meaning rather than exact keyword matches. "Find code that handles user authentication" would match functions named `verifyToken`, `checkCredentials`, and `validateSession` even though none contain the word "authentication." Powered by embeddings.

Streaming: Receiving an AI response token by token as it is generated, rather than waiting for the complete response. Essential for user-facing AI features — users see text appearing immediately rather than waiting seconds for the full response.

System Prompt: Instructions provided to the model that set its behavior for the entire conversation. In AI-powered features, the system prompt defines the AI's role, constraints, and personality. In coding tools, project configuration files (CLAUDE.md, .cursorrules) serve as system prompts.

Temperature: A parameter that controls randomness in AI output. Temperature 0 produces the most deterministic output (same input yields nearly the same output). Higher temperatures produce more varied and creative responses. For code generation, lower temperatures (0–0.3) are generally better. For brainstorming, higher temperatures (0.7–1.0) work well.

Token: The basic unit of text that AI models process. Roughly 4 characters or 0.75 words in English. Code tends to use more tokens per word because of syntax, indentation, and special characters. Token counts determine both context window usage and API costs.

Transformer: The neural network architecture behind all modern LLMs. Introduced in the 2017 paper "Attention Is All You Need." You do not need to understand transformer internals to use AI coding tools, but knowing the term helps you follow discussions about model capabilities and limitations.

Vector Database: A database optimized for storing and searching vector embeddings. pgvector (PostgreSQL extension), Pinecone, Weaviate, and Chroma are examples. Essential infrastructure for building semantic search and RAG features.

Zero-Shot: Asking a model to perform a task without providing any examples. "Translate this Python function to Go" is a zero-shot request. Most AI coding interactions are zero-shot or few-shot.

Appendix E: Resources and Further Reading

Essential Resources for AI-Assisted Development

Official Documentation

AI Coding Tool Docs: - Anthropic Claude Code: docs.anthropic.com/claude-code - GitHub Copilot: docs.github.com/copilot - Cursor: docs.cursor.com - OpenAI API: platform.openai.com/docs - Google Gemini: ai.google.dev - Amazon Q Developer: docs.aws.amazon.com/amazonq - Windsurf: docs.windsurf.com

AI API References: - Anthropic API: docs.anthropic.com - OpenAI API: platform.openai.com/docs/api-reference - Google AI SDK: ai.google.dev/docs - AWS Bedrock: docs.aws.amazon.com/bedrock

Research and Benchmarks

- **SWE-bench:** Benchmark for evaluating AI on real GitHub issues. The most realistic measure of how well AI handles actual development tasks. swebench.com
- **HumanEval:** OpenAI's code generation benchmark. Useful for comparing raw code generation ability across models.
- **LMSYS Chatbot Arena:** Community-driven model comparison through blind pairwise evaluations. Helpful for understanding real-world quality differences between models. chat.lmsys.org

Communities

- **r/ClaudeAI** — Active community for Claude users, including Claude Code workflows and tips
- **r/CursorAI** — Cursor-specific tips, workflows, and troubleshooting
- **r/ChatGPTCoding** — Code-focused ChatGPT discussions
- **GitHub Copilot Discussions** — Official community forums on GitHub
- **Hacker News** — Frequent discussions on AI coding tools (search for specific tool names)

- **Dev.to** — Developer blog posts about AI coding workflows

Newsletters and Blogs

- **Anthropic Blog** — Research updates and Claude capability announcements
- **GitHub Blog** — Copilot feature releases and research
- **Simon Willison's Weblog** — Excellent, practical coverage of AI tools for developers
- **The Pragmatic Engineer** — Gergely Orosz's coverage of AI's impact on engineering teams
- **Latent Space** — Podcast and newsletter covering AI engineering

---

Companion Titles from Finnoybu Press

This book is part of the AI for Everyone series by Finnoybu Press, designed to help real people use AI tools effectively. Every title in the series is written in the same conversational, no-nonsense style as this book.

Collection 1: Core Tools

- **ChatGPT: The Definitive Guide (2026 Edition)** — The comprehensive guide to ChatGPT for power users and professionals.
- **ChatGPT: The Power User Guide (2026 Edition)** — Advanced workflows, custom GPTs, and API integration.
- **Claude: The Definitive Guide (2026 Edition)** — Everything you need to know about Anthropic's Claude.
- **Claude Code: The Power User Guide (2026 Edition)** — Deep dive into agentic development with Claude Code.

Collection 2: Essential Platforms

- **Gemini: The Complete Guide (2026 Edition)** — Google's Gemini models, ecosystem, and integrations.
- **Gemini: The Advanced Guide (2026 Edition)** — Advanced Gemini workflows, multimodal techniques, and Google Cloud integration.

- **Microsoft Copilot: The Workplace AI Guide (2026 Edition)** — AI across the Microsoft 365 suite for everyday productivity.
- **Microsoft Copilot: The Advanced Workplace Guide (2026 Edition)** — Advanced Copilot workflows, automation, and enterprise integration.
- **Perplexity AI: The Research Playbook (2026 Edition)** — AI-powered research with real-time citations and source evaluation.
- **Perplexity AI: The Advanced Research Playbook (2026 Edition)** — Advanced research workflows, competitive intelligence, and deep analysis.

Collection 3: Authoritative Series

- **AI Systems Playbook (2026 Edition)** — Architecture, integration, and operations for AI-powered systems.
- **AI Prompt Engineering (2026 Edition)** — The complete guide to prompt engineering across models and use cases.
- **AI for Developers (2026 Edition)** — Building better software with AI assistants (this book).
- **AI for Business (2026 Edition)** — The executive's guide to AI strategy, adoption, and governance.
- **AI for Students (2026 Edition)** — The academic's guide to AI-assisted learning and research.

All titles are available in paperback and eBook formats through Amazon KDP. Visit press.finnoybu.org for the latest catalog and publication dates.

Recommended Books

AI and Machine Learning Fundamentals

- **"Designing Machine Learning Systems"** by Chip Huyen — The best practical guide to building ML systems in production. Essential if you are building AI features (Chapter 22).

- **"AI Engineering"** by Chip Huyen — Focused on building applications with foundation models. Covers RAG, fine-tuning, evaluation, and deployment.
- **"Build a Large Language Model (From Scratch)"** by Sebastian Raschka — If you want to understand what is happening inside the models you use every day.

Software Engineering

- **"Designing Data-Intensive Applications"** by Martin Kleppmann — The architecture bible. AI makes coding faster but does not replace architectural knowledge. This book builds the foundation.
- **"A Philosophy of Software Design"** by John Ousterhout — Concise and opinionated guide to managing complexity. More relevant than ever when AI generates code faster than you can understand it.
- **"The Pragmatic Programmer"** by Hunt and Thomas — Timeless principles that apply regardless of whether you or an AI is writing the code.
- **"Software Engineering at Google"** by Winters, Manshreck, and Wright — How engineering practices scale. The testing and code review chapters are especially relevant for AI-augmented teams.

Prompt Engineering and AI Communication

- **Anthropic's Prompt Engineering Guide** — Free, comprehensive, and updated regularly. The best resource for getting better results from Claude. Available at docs.anthropic.com.
- **OpenAI's Prompt Engineering Guide** — Practical tips for GPT models. Available at platform.openai.com/docs/guides/prompt-engineering.

Career and Industry

- **"The Pragmatic Engineer Newsletter"** by Gergely Orosz — Essential reading for understanding how AI is changing engineering organizations. Paid newsletter, worth every penny.
- **"Staff Engineer"** by Will Larson — The skills that matter as you advance (architecture, mentoring, technical strategy) are the same skills that become more valuable in the AI era.

Online Courses and Tutorials

- **Anthropic's Claude Courses** — Free courses on prompt engineering and building with Claude. Available at anthropic.com.
- **DeepLearning.AI Short Courses** — Andrew Ng's platform offers practical, focused courses on building with LLMs, RAG, fine-tuning, and AI agents. Most are free.
- **fast.ai** — Jeremy Howard's practical deep learning courses. The "Practical Deep Learning for Coders" course is the best way to understand the technology behind AI coding tools.

Staying Current

AI tools evolve faster than any book can track. Here is how to stay current:

1. **Follow the official changelogs** for your tools (Claude Code, Copilot, Cursor). Major features drop monthly.
2. **Read one AI newsletter** consistently. Simon Willison's blog or Latent Space are good choices.
3. **Try one new thing per month.** A new tool feature, a new prompt pattern, a new workflow. Small, consistent experimentation compounds.
4. **Talk to other developers.** Your team, your community, conferences. Shared experience is the fastest way to learn what works in practice.

The fundamentals in this book — how to prompt effectively, how to verify AI output, how to integrate AI into your workflow — will remain relevant even as specific tools and models change. Master the principles, and the tools take care of themselves.

www.ingramcontent.com/pod-product-compliance
Lightning Source LLC
LaVergne TN
LVHW060023210926
848481LV00013B/3781
* 9 7 9 8 9 5 0 5 3 3 1 6 7 *